D0922492

CHARLES DARWIN'S THE LIFE OF ERASMUS DARWIN

Charles Darwin's book about his grandfather, *The Life of Erasmus Darwin*, is curiously fascinating. Before publication in 1879, his text was shortened by 16%, with several of the cuts directed at its most provocative parts. The cutter, with Charles's permission, was his daughter Henrietta – an example of the strong hidden hand of meek-seeming Victorian women. This first unabridged edition, edited by Desmond King-Hele, includes all that Charles originally intended, the cuts being restored and printed in italics.

Erasmus Darwin was one of the leading intellectuals of the eighteenth century. He was a respected physician, a well-known poet, a keen mechanical inventor, and a founding member of the influential Lunar Society. He also possessed an amazing insight into the many branches of physical and biological science. Most notably, he adopted what we now call biological evolution as his theory of life, 65 years prior to Charles Darwin's *Origin of Species*.

DESMOND KING-HELE is the leading authority on Erasmus Darwin. A Fellow of the Royal Society since 1966, he has written and edited numerous books, including the *Letters of Erasmus Darwin*. In 1999, he was awarded the Society of Authors' Medical History Prize for his biography *Erasmus Darwin: a Life of Unequalled Achievement*.

Erasmus Darwin at the age of 38, painted by his friend and patient Joseph Wright of Derby in 1770.

Charles Darwin's
The Life of
Erasmus Darwin

First unabridged edition

Edited by

DESMOND KING-HELE

CAMBRIDGE
UNIVERSITY PRESS

AURORA PUBLIC LIBRARY

CAMBRIDGE UNIVERSITY PRESS
Cambridge, New York, Melbourne, Madrid, Cape Town,
Singapore, São Paulo, Delhi, Tokyo, Mexico City

Cambridge University Press
The Edinburgh Building, Cambridge CB2 8RU, UK

Published in the United States of America by Cambridge University Press, New York

www.cambridge.org
Information on this title: www.cambridge.org/9780521298742

Editorial material © D. King-Hele 2003

This publication is in copyright. Subject to statutory exception
and to the provisions of relevant collective licensing agreements,
no reproduction of any part may take place without the written
permission of Cambridge University Press.

Abridged editions: First edition, Murray, 1879
German translation, 1880
Second edition, Murray, 1887
Reprints: Gregg International, 1971
Chatto and Pickering, 1989
First unabridged edition published 2003
First paperback edition 2011

A catalogue record for this publication is available from the British Library

Library of Congress Cataloguing in Publication data

Darwin, Charles, 1809–1882.
The life of Erasmus Darwin / by Charles Darwin; edited by
Desmond King-Hele.–1st unabridged ed.
    p.   cm.
Originally published: 1879.
Includes bibliographical references (p.   ).
ISBN 0 521 81526 6 (hc)
1. Darwin, Erasmus, 1731–1802.   2. Naturalists–England–Biography.
3. Physicians–England–Biography.   1. King-Hele, Desmond, 1927–   11. Title.
QH31 .D3 D37   2003   508'.092–dc21   2002067073
[B]

ISBN 978-0-521-81526-0 Hardback
ISBN 978-0-521-29874-2 Paperback

Cambridge University Press has no responsibility for the persistence or
accuracy of URLs for external or third-party internet websites referred to in
this publication, and does not guarantee that any content on such websites is,
or will remain, accurate or appropriate.

Dedicated to
GEORGE PEMBER DARWIN
1928–2001
whose generous donations of family papers
to Cambridge University Library
were the inspiration for this book

Contents

Introduction

Charles Darwin's book about his grandfather Erasmus Darwin is curiously fascinating. Many people see Charles Darwin (1809–1882) as the most influential man of the last three centuries in bringing about a durable change in world-views. Indeed he was a strong candidate a few years ago for 'Man of the Millennium'.

You might therefore expect that all the books written by Charles would by now have been published in full. That is not so. His *Life of Erasmus Darwin* was shortened by 16% before publication in 1879, and several of the cuts were directed at its most provocative parts. The cutter, with Charles's permission, was his daughter Henrietta – an example of the strong hidden hand of meek-seeming Victorian women.

This first unabridged edition includes all that Charles originally intended, the cuts being restored and printed in italics.

The subject of the book, Dr Erasmus Darwin (1731–1802), has grown in stature during the twentieth century and is now seen as having achieved more in a wider variety of fields than anyone since. He was famous as a physician in the English Midlands for thirty years, and after his massive treatise on animal life, *Zoonomia*, was published in 1794, he was recognized as the leading medical author of the decade. And this happened when he was already securely in place as the leading English poet of the 1790s, or perhaps, as Coleridge said in 1797, 'the first *literary* character of Europe'. Erasmus Darwin's fame as a poet did not outlast the century; but he greatly influenced Blake, Wordsworth, Coleridge and Shelley. Earlier in his life Erasmus Darwin had been a keen and capable mechanical inventor: he devised the method of steering used in modern cars, for example. He was socially skilful too, and created several Societies, including the Lunar Society of Birmingham. His greatest talent, however, was an amazing insight in many branches of physical and biological science. For example, he was the first to explain how clouds form and, most relevantly for this book, he adopted what we now call biological evolution as his theory of life, in 1770. Many years later he risked publishing his evolutionary views, only to be rebuffed. Most people did not wish to see God deprived of his role in creating species; and everyone condemned the demeaning idea that humans had animals as distant ancestors and were no more than humanimals.

Charles's book was the first biography of Erasmus, and all subsequent biographers have been deeply indebted to it.

The book is short, direct in style, and free of all pomposity. It is also quite jovial at times, a reminder that in the Darwin household, 'the merriment, the jokes, the fun' would all be from Charles. He does not attempt a coherent narrative: instead he produces a succession of real-life pictures, more like a modern television biopic, with himself as the witty and (fairly) authoritative presenter. In his scientific books Charles was constrained by the conventions of science-writing: here (and in his *Autobiography*) he writes freely and fluently. Sometimes he teases us. Sometimes he is indignant, as when he condemns 'those bigots . . . too ignorant to be able to see their own ignorance'. He thinks he is scrupulously impartial, and free of family favouritism. Indeed Charles seems like St Peter at the celestial gate when he weighs up the pros and cons of Erasmus's moral character.

Erasmus and Charles each published a scheme of evolution, 65 years apart: it was an unbreakable bond between them. Charles habitually calls Erasmus 'my grandfather' in the book, and also when writing to other grandchildren of Erasmus, such as his sisters or cousins. He seems unaware of the significance of this possessive phrasing.

Erasmus got nowhere with his presentation of evolution: he was a century ahead of his time, as we now smugly say. It was Charles, proceeding cautiously over many years, who persuaded his contemporaries to take seriously the idea of evolution by natural selection, a world-view that has been amply vindicated in recent years.

For Charles, his strong bond with Erasmus was full of problems. If Charles praised Erasmus's evolutionary writings, people would say that Erasmus had all the ideas first, and Charles merely filled in the detail. Indeed Bishop Wilberforce, in his famous review of the *Origin of Species*, accused Charles of reviving the speculations of his 'ingenious grandsire'. So, when Charles started the book, he felt that Erasmus would have to be 'put down' rather than praised, and he took the line that Erasmus was not very important as a free-standing person. The book was being written as an item of family history, not because Erasmus's life-story needed to be told.

To Charles's great credit, he gradually changed his mind: 'the more I read of Dr. D. the higher he rises in my estimation', he wrote. This is reflected on pages 59–60, where Charles says Erasmus had 'vividness of imagination', 'great originality of thought', 'the true spirit of a philosopher' and 'uncommon powers of observation', all applied in a 'surprising' diversity of subjects. His final tribute to Erasmus (on pages 88–9) is just as generous, but was selected for deletion by Henrietta and remained unknown for a hundred years.

The book is quite a minefield of such ironies and paradoxes; some of them spring from family attitudes.

There are several paradoxes buried in Darwin–Wedgwood family history. Though you might not guess it from his many references to 'my grandfather', Charles actually had two grandfathers, Erasmus

Darwin and Josiah Wedgwood, who were close friends for thirty years and are both well known today as among the leading men of the eighteenth century. Yet Charles, and his wife Emma, who was also a grandchild of Josiah Wedgwood, knew very little about either of them and did not value their achievements. Many small pieces of inherited Wedgwood ware were damaged or destroyed when used as playthings by their children; and their two Wedgwood copies of the Portland Vase were sold to buy a billiard table for their home at Down House in Kent.

This paradox is all the sharper because the Darwins and Wedgwoods formed a close-knit extended family. Nearly everyone, it seems, knew and liked 'their sisters and their cousins and their aunts'. There were plenty of them, because the three grandfathers of Charles and Emma had 33 children, of whom 27 survived until the age of eighteen. Yet there is scarcely a black sheep in sight. Only anxiety over illness and grief over death disturbed them as they led their respectable lives.

This strong 'horizontal' family unity and lack of 'vertical' awareness arose because the two grandfathers had to work for their living, Darwin as a doctor and Wedgwood 'in trade'. Wedgwood left an immense fortune, and his sons were able to live as gentlemen. Their children, including Charles's wife Emma (daughter of Josiah Wedgwood II), were brought up in affluence. The Darwins shared in the affluence, because Charles's father Robert Darwin married Susannah Wedgwood, the wealthy eldest daughter of Josiah I. Robert managed the money well, and his children never needed to work for a living. So the Victorian Wedgwoods and Darwins lived affluently and conformed to a well-known syndrome: they preferred to forget the hard work of their grandfathers that made possible their privileged status.

The main wishes of Charles's father Robert Darwin were to be respectable, respected and rich. As a result, he sharpened several paradoxes and tensions connected with evolution.

Erasmus Darwin was much involved in 1765 with Josiah Wedgwood's promotion of the Grand Trunk Canal. In 1767 Wedgwood sent Darwin a big bag of bones found during the excavation of the Harecastle Tunnel north of Stoke. Doctors knew about bones; but not these. Though battled, Erasmus was still chirpy: 'The horn is larger than any modern horn I have measured, and must have been that of a Patagonian ox, I believe', he wrote. Wedgwood was not deceived by this teasing. The humour was only a cover-up. Erasmus was fascinated and disturbed by these fossil remains of large extinct creatures. He knew that his own father, Robert, had discovered what is today recognized as the first fossilized skeleton of a large part of a plesiosaur: it was described in the *Philosophical Transactions* of the Royal Society in 1719, and is now on display at the Natural History Museum in London.

After rattling the bones around in his mind, Erasmus soon adopted what was later called the theory of common descent, the belief that all life seen today is descended from one microscopic ancestor – a 'single living filament' as he later called it. The theory obviously implies that species change over long ages, and explains the unearthed fossils of unknown animals.

The Darwin family coat-of-arms sported three scallop shells; so Erasmus added the motto E *conchis omnia*, or 'everything from shells'. In 1770 he had the arms and motto painted on his carriage. Erasmus lived in Lichfield at the edge of the Cathedral Close, and Canon Seward of Lichfield Cathedral soon noticed the motto. He wrote a satirical poem, accusing Erasmus of 'renouncing his Creator':

> Great wizard he! by magic spells
> Can all things raise from cockle shells,

with much more in the same vein. Thus exposed, Erasmus had to paint out the motto on his carriage: he could not risk offending the rich patients on whom his livelihood depended. But he kept E *conchis omnia* on his bookplate – and presumably did not lend any books to the satirical Canon.

Erasmus eventually published his mature views on evolution (as we now call it) in 1794, tucked away near the end (pages 482–537) in Volume I of his *Zoonomia*. Here, and more explicitly in his last poem *The Temple of Nature*, he confidently expresses his view that life has developed over 'millions of ages' from microscopic specks arising spontaneously in primeval seas, through fishes and amphibians to land animals and 'humankind'. In the poem he gives vivid pictures of the struggle for existence among animals and plants, and the consequent 'survival of the fittest' (though not using these words, which came later).

In *Zoonomia* he notes how changes in the forms of animals during their lives (tadpole to frog, etc.) show that change is rife in nature. He discusses the effects of artificial selection in modifying species, noting that monstrosities (mutations) are often inherited. He proposes a theory of heredity in terms of 'fibrils or molecules' from male or female, which combine to produce the new embryo. 'Lust, hunger and security', he says, are the controlling forces of change: animals become adapted to their food supply, for example the varied beaks of finches; mimicry and protective coloration are important. He defines sexual selection: in species where the males 'combat each other' for 'exclusive possession of the females', the outcome is, he says, 'that the strongest and most active animal should propagate the species, which should thence become improved'.

Charles Darwin read *Zoonomia* when he was sixteen or seventeen, and also listened to a panegyric in praise of evolution from his friend Dr Robert Grant at Edinburgh University. 'At this time I greatly admired the *Zoonomia*', he says. But neither Grant nor *Zoonomia* had 'any effect on my mind'. This is true: otherwise he would have become an evolutionist before going on the voyage of the *Beagle*, rather than after. There is a major paradox here, perhaps best explained by assuming that the voyage quite transformed him.

Several ironies hover round Dr Robert Darwin, the son of Erasmus and the father of Charles – sometimes called the missing link between them. Robert was dominated by his father, who organized his education so efficiently that Robert was practising as a physician in Shrewsbury at the age of twenty; and he was immediately successful, even though he never wanted to be a doctor. Robert also found himself elected a Fellow of the Royal Society at the age of 21.

Robert reacted against his father by refusing to do any more science and by seeking to lead a life of privacy and respectability, rather than of public exposure. In this he was constant, consistent and conservative throughout his adult life. But by then he already held many fixed ideas imbibed from Erasmus. Robert, like his father, would gain the confidence of patients by sympathy, careful observation and, if appropriate, boundless optimism. Robert abstained from alcohol, as Erasmus had done. Robert was sceptical of religion, and was an undeclared unbeliever, according to Charles. Robert also adopted his father's belief in evolution, and seems to have done so quite enthusiastically: Robert was not an activist, and yet he took the trouble to have a bookplate made with the dreaded motto E conchis omnia. It can be seen in many of his books now in Cambridge University Library.

When Robert was 32, already married to Susannah and quite set in his ways, the official backlash to Zoonomia struck. The war against the French was going badly in 1797, with Napoleon rampant in Europe and the British Navy in mutiny. A new magazine was begun in 1798, to boost morale and combat all ideas subversive of the established order. Called The Anti-Jacobin, it was controlled by George Canning, a junior government minister and later Prime Minister. With two collaborators, Canning set out to destroy Erasmus's reputation with a poem called The Loves of the Triangles, in parody of Erasmus's poem The Loves of the Plants, and backed up by long notes ridiculing Erasmus's ideas, particularly the absurd notion that human beings evolved from lower forms of life. This onslaught pushed Erasmus off his pedestal: his status as the leading poet gradually crumbled, and his evolutionary theory was little heeded. Robert, horrified at this public brawl,

went underground with his evolutionary beliefs. There is no in-
dication that he renounced evolution: he was set in his ways, and
he did bring up his children in an evolution-friendly atmosphere,
though he probably never discussed the subject with them.

So, when Charles set off on the voyage of the *Beagle* in 1831,
he had not yet arrived at evolution, although his grandfather and
father were evolutionists.

This paradox leads to another. Robert was a formidable figure
by the 1830s, physically and psychologically. He was 6 feet 2 inches
tall and weighed more than 24 stone (152 kg). Charles said 'he
was the largest man I ever saw'. And when Robert was in a room,
all talk had to be directed to him. Charles was dominated by him:
'his reverence for him was boundless and most touching', as his
son Francis remarked. Yet Charles probably never knew of his
father's suppressed belief in evolution.

It was primarily his experiences on the voyage of the *Beagle*
that made Charles turn towards evolution on his return in 1836.
However, when he began his first notebook on the species question, he entitled the notebook 'Zoonomia', and twice referred to
Erasmus's book.

In 1839 Charles wrote to his second cousin William Darwin
Fox and asked him, 'Can you tell me from *memory* what the motto
to our crest is, for I mean to have a seal solemnly engraved'. So
Charles was unaware of E *conchis omnia*, the motto that adorned
so many of his father's books, and also those he inherited from
Erasmus. Did Charles forget? Did he never see his father's books?
The reply from Fox would only have misled him, because Fox was
the grandson of Erasmus's elder brother William, and would only
have known the motto of the elder branch of the family, *Cave et
aude*, 'take care and be bold'.

Piling paradox on paradox becomes exhausting, and I shall
cease doing so. Many more can be found in Ralph Colp's richly
referenced paper on 'The relationship of Charles Darwin to
the ideas of his grandfather, Dr Erasmus Darwin', in *Biography*,
Vol.9, pp.1–24 (1986). Further details of the facts, and references

for the quotations, can be found in Colp's paper and in my biography *Erasmus Darwin: a Life of Unequalled Achievement* (London: de la Mare, 1999), particularly pages 78, 89–91, 297–301, 358–9 and 363–8. A beguiling panorama of the Darwin–Wedgwood extended family emerges from Edna Healey's book *Emma Darwin* (London, Headline, 2001; paperback, 2002).

The writing of Charles's *Life* of Erasmus was a remarkably rapid international enterprise. All the action occurred in 1879, twenty years after the publication of the *Origin of Species*. By now Charles had achieved world renown for his work on biological evolution by natural selection. His 70th birthday, on 12 February 1879, was celebrated in a special issue of the German science journal *Kosmos*, and the final article, by Ernst Krause, was on 'Erasmus Darwin, the grandfather and forerunner of Charles Darwin'. On 9 March Charles wrote to Krause, offering to have this 28-page article translated into English by William Dallas. In reply, Krause politely offered to enlarge his essay. Charles agreed, and said he would write a 'preface'. He next asked several cousins what they knew about Erasmus. They responded well; and Charles himself then found he already had some boxes with letters written by Erasmus.

Charles began writing his 'preface' in mid-May and, boosted by the new materials, it grew into a 'Preliminary Notice' of about 100 pages, which he finished early in June. He sent it for printing, and in July received proofs, which he showed to a number of relatives. Several of them were favourably impressed, but his daughter Henrietta, who had helped him with previous books, thought it 'dull' and 'too long'. His son Leonard was also critical and suggested that Henrietta should cut up the proofs, rearrange the pieces, and reduce the length.

Disappointed by these reactions – 'never again will I be tempted out of my proper work' – Charles returned to his scientific research and left Henrietta to do the cutting up and cutting down.

Meanwhile Dallas finished translating Krause's enlarged essay in August, and John Murray agreed to publish the book. It came out in November.

Charles's 'Preliminary Notice' filled 129 small pages and was 50% longer than Krause's 86-page essay. As Charles had only meant to write a preface, the title-page reads: 'Erasmus Darwin, by Ernst Krause'; then, lower down and in smaller type, 'With a Preliminary Notice by Charles Darwin'. In 1887, after Charles's death, a second edition was brought out by his son Francis, who added a synopsis and changed the title-page to: 'The Life of Erasmus Darwin by Charles Darwin'; Krause is reduced to smaller type.

My bare summary of the events of 1879 has left Henrietta with rather a negative role. This is unfair to her. It was at Charles's request that she cut up and cut down the text: he had been very grateful for her help with *The Descent of Man*, and trusted her to make improvements. Most of her cuts are well chosen, though some have to be called censorship; and her stylistic changes are nearly all for the better. She was later very successful in editing *Emma Darwin: a Century of Family Letters* (1904).

The paradoxes that haunt this Introduction now re-emerge for a final fling: 'The Life of Erasmus Darwin, by Charles Darwin' does not appear in its own right in the thousand-volume *National Union Catalog*, and the book is very often omitted altogether in bibliographies of Charles Darwin. Such omissions are insulting to Charles, who worked hard on the book. His known surviving correspondence for 1879 includes 189 letters concerned with the book. What a paradox *par excellence* if this present edition is itself doomed to be catalogued under 'Krause'!

The Text of the Present Unabridged Edition

This edition is based on the corrected first proofs in the Darwin Archive at Cambridge University Library (DAR 210.11:45), and uses the title of the second edition (1887). I have included about three pages of additional material inserted in the 1879 book with Charles's approval.

My aim has been to make the text as clean and readable as possible. Henrietta's deletions have been restored and *printed in italics whenever they exceed five words. I have silently accepted most of* Henrietta's stylistic changes, but have not adopted her rearrangement of the text, because the original arrangement seems more logical.

Charles's numerous footnotes are of two distinct types, which I have treated differently. Notes of type 1 are narrative text that adds to the story. I have inserted these at suitable places in the main text, *placing them within curly brackets:* {...}. Notes of type 2 merely give the sources of quotations. For these I have inserted a superscript number at the appropriate place in the main text, the content of the note being printed in the List of Charles Darwin's References at the end.

Fortunately, the titles of books are not italicized in the proofs or the 1879 book, but are set within single quotes, e.g. 'The Botanic Garden'. This style has been preserved, as have the notation for dates and several other standard conventions of the time, such as Mr. (with a dot). In the proofs and the book, most quotations are set between double quotes: "...". This style is preserved for quotations of fewer than five lines: longer quotations are inset instead.

I have corrected Charles Darwin's factual errors, and have recorded, within square brackets immediately afterwards, the original erroneous text, preceded by the words 'C.D. has'. For example:

Dr. Darwin married the widow of Colonel Edward Pole [C.D. has 'Colonel Chandos Pole'] of Radburn Hall....

After the text and References come my quite lengthy 'Notes on the Text'. They include: explanations of C.D.'s errors mentioned above; sources, manuscript or published, for C.D.'s many unreferenced quotations; brief identifications of about 130 people named; information to fill gaps or obscurities in C.D.'s episodic presentation; and comments on curiosities in the text.

After the Notes there is a Chronology of Erasmus Darwin's life (Appendix A); a selective Darwin family tree (Appendix B); a list of selected relevant books and papers (Appendix C); an outline of Krause's essay (Appendix D); and a table linking the page numbers in the 1879 book with those in the proofs – an editor's nightmare unravelled (Appendix E).

If we exclude the long quotations in small type (from letters or other books), Charles's original text runs to 1644 lines and the deletions total 266 lines, or 16%.

THE LIFE OF ERASMUS DARWIN
BY CHARLES DARWIN

SYNOPSIS

[supplied by the editor: the original text is devoid of subject headings]

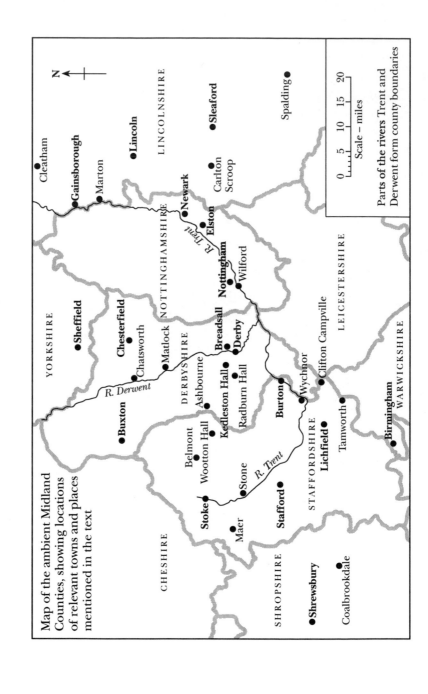

Map of the ambient Midland Counties, showing locations of relevant towns and places mentioned in the text

Parts of the rivers Trent and Derwent form county boundaries

Scale – miles

0 5 10 15 20

Preface

BY CHARLES DARWIN

In the February number, 1879, of a well-known German scientific journal, 'Kosmos', Dr. Ernst Krause published a sketch of the life of Erasmus Darwin, the author of the 'Zoonomia', 'Botanic Garden', and other works. This article bears the title of a 'Contribution to the history of the Descent-Theory'; and Dr. Krause has kindly permitted my brother and myself to have a translation made of it and published. {Mr. Dallas has undertaken the translation, and his scientific reputation, together with his knowledge of German, is a guarantee for its accuracy.}

As I have private materials for adding to the knowledge of Erasmus Darwin's character, I have written a preliminary notice. These materials consist of a large collection of letters written by him; of his common-place book in folio, in the possession of his grandson Reginald Darwin; of some notes made by my father shortly after the death of his father, together with what little I can clearly remember that my father said about him; also some statements by his daughter, Violetta Darwin, written down at the time by her daughters, the Miss Galtons, and various published notices. To them must be added the 'Memoirs of the Life of Dr. Darwin', by Miss Seward, which appeared in 1804, and a lecture by Dr. Dowson on 'Erasmus Darwin, Philosopher, Poet, and Physician', published in 1861, which contains many useful references and remarks.

It has been the fate of many celebrated men with strongly-marked characters to have been grossly calumniated; and few have suffered more in this respect than Erasmus Darwin. The publication of the present essay seemed to me a good opportunity for showing how utterly groundless most of these

5

calumnies were. I hope also to be able to give a truer and fuller, yet far from complete, idea of his general character than has yet appeared.

Dr. Krause has taken great pains, and has added largely to his essay as it appeared in 'Kosmos'; and my preliminary notice, having been written before I had seen the additions, unfortunately contains much repetition of what Dr. Krause has said. In fact the present volume contains two distinct biographies, of which I have no doubt that by Dr. Krause is much the best. I have left it almost wholly to him to treat of what Dr. Darwin has done in science, more especially in regard to evolution.

{Since the publication of Dr. Krause's article, Mr. Butler's work, 'Evolution, Old and New, 1879', has appeared, and this includes an account of Dr. Darwin's life, *without anything new having been added,* and of his views on evolution.}

Note that here, and in pages 7–91, text deleted from the 1879 book is printed in italics. For notes on the text, see pages 97–134.

The Life of Erasmus Darwin

As the character of a man depends in some degree on the circumstances under which he has been brought up, it will be advisable to give a very short account of the family to which Erasmus Darwin belonged. It is more important to show to what extent a man inherits and transmits his characteristic qualities; for every addition, however small, to our knowledge on this head is a public benefit, as well as spreading a belief in inheritance.

{As the name Darwin is an unusual one, I may mention that in the Cottonian Library, now in the British Museum, there is a large and very rare book, on the History of Lichfield; and in this book the antiquary, Sir R. Cotton, who was born in 1570 and died in 1631, made the following entry: "This Booke was found in the thatch of an House at Clifton-Campville, in the demolishinge thereof. And was brought to mee by Mr. Darwin". Clifton-Campville is near Tamworth, in Staffordshire. Whether the Mr. Darwin who made this discovery was a member of the family we do not know.}

Erasmus Darwin was descended from a family of yeomen who lived for several generations on their own land, apparently of considerable extent, at Marton in Lincolnshire. The great-grandson of the first Darwin of whom we know anything seems to have been a loyal man, for in his will made in 1584 he [Richard Darwin] bequeathed 3s. 4d. "towards the settynge up of the Queene's Majesties armes over the queare [choir] door in the parishe Churche of Marton".

His son William, born about 1575, possessed a small estate at Cleatham, at no great distance from Marton. He considered himself a gentleman, bore arms and married a lady. He was also yeoman of the armoury of Greenwich to James I and Charles I. This office was probably almost a sinecure, and certainly of very small value.

{The greater part of the estate of Cleatham was sold in 1760. A cottage with thick walls, some fish-ponds and old trees, alone show where the "Old Hall" once stood. A field is still called the "Darwin Charity", from being subject to a charge, made by a later Mrs. Darwin, for buying gowns for four old widows every year.}

William Darwin died in 1644, and we have reason to believe from gout. It is, therefore, probable that Erasmus, as well as many other members of the family, inherited from this William, or some of his predecessors, their strong tendency to gout; and it was an early attack of gout which made Erasmus a vehement advocate for temperance throughout his whole life.

The second William Darwin (born 1620) served as Captain-Lieutenant in Sir W. Pelham's troop of horse, and fought for the king. His estate was sequestrated by the Parliament, but he was afterwards pardoned on payment of a heavy fine. In a petition to Charles II he speaks of his almost utter ruin from having adhered to the royal cause, and it appears that he had become a barrister. This circumstance probably led to his marrying the daughter of Erasmus Earle, Serjeant-at-law, *who lived at Heydon Hall in Norfolk and represented Norwich in the Long Parliament*; hence probably Erasmus Darwin derived his Christian name.

The eldest son from this marriage, William (born 1655), married the heiress of Robert Waring, of Wilford, a family of much consideration in Nottinghamshire. This lady also inherited, by an indirect course, the manor of Elston, which has remained ever since in the family; and Erasmus, the subject of the present notice, was born at Elston Hall.

The William Darwin who married the heiress is said to have been a good workman, and he may have transmitted his mechanical taste to his grandson, Erasmus. I possess an ivory box made by him, prettily ornamented on one side, with his crest on the other side. There is a portrait of him at Elston, showing a pleasing and refined countenance.

Elston Hall, where Erasmus Darwin was born, as it existed before 1754. From a drawing by Violetta H. Darwin.

The third William Darwin had two sons – William, and Robert, who was educated as a barrister and was the father of Erasmus. I suppose the Cleatham and the Waring properties were left to William, who seems to have followed no profession, and the Elston estate to Robert; for when the latter married he gave up his profession, and lived ever afterwards at Elston.

There is a portrait of Robert at Elston Hall, and he looks, with his great wig and bands, like a dignified doctor of divinity. He seems to have had some taste for science, for he was an early member of the famous Spalding Club; and the celebrated antiquary, Dr. Stukeley, in 'An account of the almost entire Sceleton of a large animal, &c.', published in the 'Philosophical Transactions', April and May, 1719, begins his paper as follows:— "Having an account from my friend, Robert Darwin, Esq., of Lincoln's Inn, a Person of Curiosity, of a human Sceleton (as it was then thought) impressed in Stone, found lately by the Rector of Elston, &c." Stukeley then speaks of it as a great rarity, "the like whereof has not been observed before in this island, to my knowledge".

Erasmus wrote to his friend, Dr. Okes, with much frankness about his father's character, shortly after Robert's death in 1754:

> He was a man of more sense than learning; of very great industry in the law, even after he had no business, nor expectation of any. He was frugal, but not covetous; very tender to his children, but still kept them at an awful kind of distance. He passed through this life with honesty and industry, and brought up seven healthy children to follow his example.

Judging from a sort of litany written by him, and handed down in the family, Robert was a strong advocate of temperance, which his son ever afterwards so strongly advocated:

> From a morning that doth shine,
> From a boy that drinketh wine,
> From a wife that talketh Latine,
> Good Lord deliver me.

It is suspected that the third line may be accounted for by his wife having been a very learned lady.

The eldest son of Robert, christened Robert Waring, succeeded to the estate of Elston, and died there at the age of ninety-two, a bachelor. He had a strong taste for poetry, like his youngest brother Erasmus, *as I infer from the latter having dedicated a MS. volume of juvenile poems to him, with the words, "By whose example and encouragement my mind was directed to the study of poetry in my very early years". The two brothers also corresponded together in verse.* Robert likewise cultivated botany, agreeing also in this respect with Erasmus, and when a rather old man he published his 'Principia Botanica'. This book in MS. was beautifully written, and my father declared that he believed it was published because his old uncle could not endure that such fine calligraphy should be wasted. But my father was hardly just, as the work contains many curious notes on biology—a subject wholly neglected in England in the last century. The public, moreover, appreciated the book, as the copy in my possession is the third edition.

The second son, William Alvey, *became the ancestor of the elder branch of the family, the present possessors of Elston Hall.*

A third son, John, became the rector of Elston, the living being in the gift of the family; and of him I have heard nothing worthy of notice.

The fourth son, and the youngest of the children, was Erasmus, the subject of the present memoir, who was born on the 12th December 1731, at Elston Hall.

———————————— ⤳ ————————————

Before proceeding to give some account of his life and character, it may be well to see how far he transmitted his characteristic qualities to his children. He had three sons by his first wife (besides two children who died in infancy).

His eldest son, Charles (born September 3, 1758), was a young man of extraordinary promise, but died (May 15, 1778) before he

was twenty-one years old from the effects of a wound whilst dis-
secting the brain of a child. He inherited from his father a strong
taste for various branches of science, for writing verses, and for
mechanics. "Tools were his playthings", and making "machines
was one of the first efforts of his ingenuity, and one of the first
sources of his amusement".[1]

He also stammered like his father, who sent him to France
when between eight and nine years old (1766–67), with a private tutor, as he thought that if Charles was not allowed to speak
English for a time the habit of stammering might be lost; and
it is a curious fact (as I hear from one of Dr. Darwin's granddaughters) that in after years when speaking French he never
stammered. At a very early age he collected specimens of all
kinds. When sixteen years old he went [C.D. has 'was sent'] for a
year to Oxford, but he did not like the place, and thought (in the
words of his father) "that the vigour of the mind languished
in the pursuit of classical elegance, like Hercules at the distaff,
and sighed to be removed to the robuster exercise of the medical
school of Edinburgh".

He stayed three years at Edinburgh, working hard at his medical studies, and attending "with diligence all the sick poor of
the parish of Waterleith, and supplying them with the necessary
medicines". The Æsculapian Society awarded him its first gold
medal for an experimental enquiry on pus and mucus. Several
notices of him appeared in various journals; and all the writers
agree about his uncommon energy and abilities.

He seems also, like his father, to have excited the warm affection of his friends. The venerable Professor Andrew Duncan,
who had Charles buried in his own family vault, spoke to me
about him with the warmest affection forty-seven years after his
death. Professor Duncan cut a lock of hair from the corpse, and
took it to a jeweller, whose apprentice, afterwards the famous
Sir H. Raeburn, set it in a locket for a memorial.[2] The inscription on his tomb, written by Erasmus, says, with more truth than
is usual on such occasions: "Possessed of uncommon abilities

and activity, he had acquired knowledge in every department of medical and philosophical science, much beyond his years".

Dr. Darwin was able to reach Edinburgh before Charles died, and had at first hopes of his recovery; but these hopes, as he informed his second son [C.D. has 'my father'] "with anguish", soon disappeared. Two days afterwards he wrote to Josiah Wedgwood to the same effect, ending his letter with the words, "God bless you, my dear friend, may your children succeed better". Two and a half years afterwards he again wrote to Wedgwood, "I am rather in a situation to demand than to administer consolation".

About the character of his second son, Erasmus (born 1759), I have little to say, for, though he wrote poetry, he seems to have had none of the other tastes of his father. He had, however, his own peculiar tastes, viz. genealogy, the collecting of coins, and statistics. When a boy he counted, as I was told by my father, all the houses in the city of Lichfield, and found out the number of inhabitants in as many as he could; he thus made a census, and when a real one was first made, his estimate was found to be nearly accurate. His disposition was quiet and retiring.

My father had a very high opinion of his abilities, and this was probably just, for he would not otherwise have been invited to travel with, and pay long visits to, men so distinguished in different ways as Boulton, the engineer, and Day, the moralist and novelist. He was certainly very ingenious, and he detected by a singularly subtle plan the author of a long series of anonymous letters, which had caused during six or seven years, extreme annoyance and even misery to many of the inhabitants of the county; and the author was found to be a county gentleman of not inconsiderable standing.

Erasmus practised as a solicitor in Derby [C.D. has 'Lichfield'] and was so successful that when only forty years old he thought of retiring from business and building a cottage on his father's former botanic garden. His father considered this a great mistake, and in a letter to my father (August 8, 1799) says: "all which I much disapprove. Therefore you will please not to mention it, and I hope it will fall through." He did, however, soon afterwards purchase Breadsall Priory, near Derby, with the intention of soon retiring to a life of tranquillity. But no such quiet end was in store for him, and his unhappy fate will afterwards be related.

The third son, Robert Waring Darwin (my father, born 1766), did not inherit from his father any aptitude for poetry or mechanics, nor did he possess, as I think, a scientific mind, though he had a strong taste for flowers and gardening. He published in vol. lxxvi of the 'Philosophical Transactions' a paper on Ocular Spectra, which Wheatstone told me was a remarkable production for the period; but I believe that he was largely aided in writing it by his father. He was elected a Fellow of the Royal Society in 1788. I cannot analyse why my father's mind did not appear to me fitted for advancing science; for he was fond of theorising, and was incomparably the most acute observer whom I ever knew. But his powers of observation were all turned to the practice of medicine, and still more closely to human character. *Here would not be the proper place to show with what extraordinary acuteness he* intuitively recognised the disposition or character, and even read the thoughts, of those with whom he came into contact. This skill partly accounts for his great success as a physician, for it impressed his patients with confidence; and my father used to say that the art of gaining confidence was the chief element in a doctor's worldly success.

His brother Erasmus [C.D. has 'father'] brought him to Shrewsbury before he was twenty-one years old; his father had given [C.D. has 'and left'] him £20, saying, "Let me know when you want more, and I will send it you". His uncle John, the rector of Elston, afterwards sent him (as I gather from an old letter) £20, and this was the sole pecuniary aid which my father ever received. {It appears however from papers in the possession of Mr Reginald Darwin that he got £1000 under his mother's settlement, and £400 from his aunt Susannah Darwin.}

I have heard my father say that his practice during the first year allowed him to keep two horses and a man-servant. Erasmus tells Mr. Edgeworth that his son Robert, after being settled in

Shrewsbury for only six months, "already had between forty and fifty patients". By the second year he was in considerable, and ever afterwards in very large practice. *I remember his coming into the room and saying, "This day, sixty years ago, I received my first fee in Shrewsbury", and he continued to practice for one or two years longer. It may be doubted whether any doctor will ever again have a considerable number of patients before arriving at the age of twenty-one years.* His success was the more remarkable, as he for some time detested the profession, and declared that if he had been sure of gaining £100 a year in any other way he would never have practised as a doctor.

He had an extraordinary memory for the dates of certain events, so that he knew the day of the birth, marriage and death of most of the gentlemen of Shropshire. This power, far from giving him any pleasure, annoyed him, for when he once heard a date it was fixed for ever in his mind. He told me that it added to his regret for the death of old friends. His spirits were generally high, and he was a great talker, but he was of an extremely sensitive nature, so that whatever annoyed or pained him, did so to an extreme degree. He was also rather easily roused to anger.

On my asking him, when too old to walk, why he did not sometimes drive out in his carriage, he answered that every road out of Shrewsbury was associated in his mind with some painful event; and so vivid was his memory that I have no doubt such events returned with all the freshness of reality. He strongly disliked extravagance, but was highly generous. A manufacturer, a friend of his, called on him and stated that he should be ruined unless he could borrow £10,000, and that he could offer only personal security for this large sum. My father, from his wonderful insight into character, was convinced of the accuracy of every word which his friend said, and lent him the money, at a time when the amount must have been a serious matter to him. It was repaid, and the manufacturer was saved from ruin.

One of his golden rules was never to become the friend of any one whom you could not thoroughly respect, and I think he acted on it. But of all his characteristic qualities, his sympathy was pre-eminent, and I believe it was this which made him for a time hate his profession, as it constantly brought suffering before his eyes.

Sympathy with the joy of others is a much rarer endowment than sympathy with their pains, and it is no exaggeration to say that to give pleasure to others was to my father an intense pleasure. He died November 13th, 1848 [C.D. has 1849]. A short notice of his life appeared in No. 74 of the 'Proceedings of the Royal Society'.

As we have been here considering how much or how little the same tastes and disposition prevail in the same family, I may be permitted to add that from my earliest days I had the strongest desire to collect objects of natural history; and this was certainly innate or spontaneous, being probably inherited from my grandfather. Some of my sons have also exhibited an apparently innate taste for science. Members of the family have been elected Fellows of the Royal Society for four successive generations.

Of the children of Erasmus by his second marriage (four sons and three daughters), one son became a cavalry officer, a second rector of Elston, and a third, Francis (born 1786, died 1859), a physician, who travelled far in countries rarely visited in those days. He showed his taste for natural history by being fond of keeping a number of wild and curious animals. One of his sons, Captain Darwin, is a great sportsman, and has published a little book, the 'Gamekeeper's Manual' (4th ed. 1863), which shows keen observation and knowledge of the habits of various animals. The eldest daughter of Erasmus, Violetta, married S. Tertius Galton, and I feel sure that their son, Francis Galton,[3] will be willing to attribute the remarkable originality of his mind in large part to inheritance from his maternal grandfather.

We may now return to Erasmus Darwin. His elder brother Robert states, in a letter to my father (May 19, 1802) that Erasmus "was always fond of poetry. He was also always fond of mechanicks. I remember him when very young making an ingenious alarum for his watch {clock?}; he used also to show little experiments in electricity with a rude apparatus he then invented with a bottle."

The same tastes, therefore, appeared very early in his life which prevailed to the day of his death. "He had always a dislike to much exercise and rural diversions, and it was with great difficulty that we could ever persuade him to accompany us."

But such was not invariably the case, for Robert celebrated, in the following doggerel verses, the fact of Erasmus, when nine years old, and of his brother, J. D., the future rector of Elston, catching a hare. Robert, who was famous afterwards in the family for his beautiful handwriting, and who was in all respects a most precise old gentleman, would have been shocked if he could have seen his own handwriting, bad grammar, and extraordinary orthography when sixteen years old. He would no doubt have declared it was quite beneath the dignity of biography to publish such verses.

A new Song in the praise of two young Hunters, 1740

One morning this winter from school J. D. came,
And him and his brother Erasmus went out to kill game,
And as it happened, which was very rare,
With hounds and 2 Spaniels killed a fine hair.

2.

Into the whome close they went to get crabs for a tart,
They no sooner got them, but a Hare they did start;
And then they run her with hounds and with horn,
And catched her before she got to the corn.

3.

One of the dogs catched her by the head,
Which made Erasmus Darwin cry war dead war dead,
But John Darwin the dogs he could not hear
Because he only cried out war, war, war.

4.

Then the sport was done and all at an end.
They brought her home and told the news to their friends.
Soon after they pouched her and stue'd her in her blood,
And everybody that eat of her, said she was good.

When ten years old (1741), Erasmus was sent to Chesterfield School, where he remained for nine years. His sister, Susannah, wrote to him at school in 1748 [old style: 1749 new style], and I give part of the letter as a curiosity, considering that she was then a young lady between eighteen and nineteen years old. She died unmarried, and my father, who was deeply attached to her, always spoke of her as the very pattern of an old lady, so nice looking, so gentle, kind, and charitable, and passionately fond of flowers. The first part of her letter consists of gossip and family news, and is not worth giving. Erasmus was sixteen years old when he answered her.

SUSANNAH DARWIN TO ERASMUS

Dear Brother,

I come now to the chief design of my Letter and that is to acquaint you with my Abstinence this Lent, which you will find on the other side, it being a strict account of the first 5 days, and all the rest has been conformable thereto; I shall be glad to hear from you with an account of your temperance this lent, which I expect far exceeds mine. As soon as we kill our hog I intend to take part thereof with the Family, for I'm informed by a learned Divine that Hogs Flesh is Fish, and has been so ever since the Devil entered into them and they ran into the Sea; if you and the rest of the Casuists in your neighbourhood are of the same oppinion, it will be a greater satisfaction to me, in resolving so knotty a point of Conscience. This being all at present I conclude with all our dues to you and Brother.

Your affectionate sister,
S. Darwin

A DIARY IN LENT

Elston, Feb. 20, 1748

Feb^ry 8 Wednesday Morning a little before seven I got up; said my Prayers; worked till eight; then took a walk, came in again and eate a farthing Loaf, then dress'd me, red a Chapter in the Bible, and spun till

One, then dined temperately viz: on Puddin, Bread and Cheese; spun again till Fore, took a walk, then spun till half an hour past Five; eat an Apple, Chatted round the Fire; and at Seven a little boyl'd Milk; and then (takeing my leave of Cards the night before) spun till nine; drank a Glass of Wine for the Stomack sake; and at Ten retired into my Chamber to Prayers; drew up my Clock and set my Larum betwixt Six and Seven.

Thursday call'd up to Prayers, by my Larum; spun till Eight, collected the Hen's Eggs; breakfasted on Oat Cake, and Balm Tea; then dress'd and spun till One, Pease Porrage Pottatoes and Apple Pye; then turned over a few pages in Scribelerus; eat an Apple and got to my work; at Seven got Apple Pye and Milk, half an hour after eight red in the Tatlar and at Ten withdrew to Prayers; slept sound; rose before Seven; eat a Pear; breakfast a quarter past Eight; fed the Cats, went to Church; at One Pease Porrage Puddin Bread and Cheese, Fore Mrs. Chappells came, Five drank Tea; Six eat half an Apple; Seven a Porrenge of Boyl'd Milk; red in the Tatlar; at Eight a Glass of Punch; filled up the vacancies of the day with work as before.

Saturday Clock being too slow lay rather longar than usal; said my Prayers; and breakfasted at Eight; at One broth Pudding Brocoli and Eggs and Apple Pye; at Five an Apple; seven Apple Pye Bread and Butter; at Nine a Glass of Wine; at Ten Prayers.

Sunday breakfast at Eight; at Ten went to the Chappell; 12 Dumplin, red Herring, Bread and Cheese; two to the Church; read a Lent Sermon at Six, and at Seven Appel Pye Bread and Butter.

Excuse hast being very cold.

ERASMUS TO SUSANNAH DARWIN

Dear Sister,

I receiv'd yours about a fortnight after the date that I must begg to be excused for not answering it sooner: besides I have some substantial Reasons, as having a mind to see Lent almost expired before I would vouch for my Abstinence throughout the whole: and not having had a convenient oppertunity to consult a Synod of my learned friends about your ingenious Conscience, and I must inform you we unanimously agree in the Opinion of the Learned Divine you mention, that Swine may indeed be fish but then they are a devillish sort of fish; and we can prove from the same Authority that all fish is flesh whence we affirm Porck not only to be flesh but a devillish Sort of flesh; and I would advise you for Conscience sake altogether to abstain from tasting it; as I can assure You I have done, tho' roast Pork has come to Table several Times; and for my own part have lived upon Puding, milk, and vegetables all this Lent; but don't mistake me, I don't mean I have not touch'd roast beef, mutton, veal, goose, fowl, &c. for what are all these? All flesh is grass! Was I to give you a journal of a Week it would be stuft so full of Greek and Latin as translation, Verses, themes, annotation, Exercise and the like, it would not only be very tedious and insipid but perfectly unintelligible to any but Scholboys.

I fancy you forgot in Yours to inform me that your Cheek was quite settled by your Temperance, but however I can easily suppose it. For the temperate enjoy an ever-blooming Health free from all the Infections and disorders luxurious mortals are subject to; the whimsical Tribe of Phisitians cheated of their fees may sit down in penury and Want, they may curse mankind and imprecate the Gods and call down that parent of all Deseases, luxury, to infest Mankind, luxury more distructive than the Sharpest Famine; tho' all the Distempers that ever Satan inflicted upon Job hover over the intemperate; they would play harmless round our Heads, nor dare to touch a single Hair. We should not meet those pale thin and haggard countenances which every day present themselves to us. No doubt men would still live their Hunderd, and Methusalem would lose his Character; fever banished from our Streets, limping Gout would fly the land, and Sedentary Stone would vanish into oblivion and death himself be slain.

I could for ever rail against Luxury, and for ever panegyrize upon abstinence, had I not already encroach'd too far upon your Patience; but it being Lent the exercise of that Christian virtue may not be amiss so I shall proceed a little furder—

{The remainder of the letter is hardly legible and unintelligible, with no signature.}

 P.S.—Excuse Hast, supper being called, very Hungry.

———————————— ⁓ ————————————

Judging from two letters—the first written in 1749 to one of his school-friends [C.D. has 'under-masters'] during the holidays, and the other to the head-master, shortly after arriving at the University of Cambridge, in 1750—he seems to have felt a degree of respect, gratitude, and affection for several masters unusual in a schoolboy. Both these letters were accompanied by an inevitable copy of verses, those addressed to the head-master being of considerable length, and in imitation of the 5th Satire of Persius.

His elder brother John [C.D. has 'His two elder brothers'] accompanied him to St. John's College, Cambridge; and this seems to have been a severe strain on their father's income. They appear, in consequence, to have been thrifty and honourably economical; so much so that they mended their own clothes; and, many years afterwards, Erasmus boasted to his second wife that, if she cut the heel out of a stocking, he would put a new one in without missing a stitch. He won the Exeter Scholarship at St. John's, which was worth only £16 per annum.

No doubt he continued to study the classics whilst at Cambridge, for he did so to the end of his life, as shown by the many quotations in his latest work, 'The Temple of Nature'. He must likewise have studied mathematics to a certain extent, for, when he took his Bachelor of Arts degree, in 1754, he [actually his brother] was at the head of the Junior Optimes.

Nor did he neglect medicine; and he left Cambridge during one term to attend Hunter's lectures in London. As a matter of course, he wrote poetry whilst at Cambridge, and a poem on 'The Death of Prince Frederick', in 1751, was published many years afterwards, in 1795, in the European Magazine.

In the autumn of 1753 [C.D. has '1754'] he went to Edinburgh to study medicine, and while there seems to have been as rigidly economical as at Cambridge; for amongst his papers there is a receipt for his board from July 13th to October 13th [1754], amounting to only £6 12s.

Mr Keir, afterwards a distinguished chemist, was at Edinburgh with him, and after his death wrote to my father (May 12, 1802):

> The classical and literary attainments which he had acquired at Cambridge gave him, when he came to Edinburgh, together with his poetical talents and ready wit, a distinguished superiority among the students there. Every one of the above-mentioned Professors {those whose lectures he attended}, excepting Dr. Whytt, had been a pupil of the celebrated Boerhaave, whose doctrines were implicitly adopted. It would be curious to know (but he alone could have told us) the progress of your father's mind from the narrow Boerhaavian system, in which man was considered as an hydraulic machine whose pipes were filled with fluid susceptible of chemical fermentations, while the pipes themselves were liable to stoppages or obstructions (to which obstructions and fermentations all diseases were imputed), to the more enlarged consideration of man as a *living being*, which affects the phenomena of health and disease more than his merely mechanical and chemical properties. It is true that about the same time, Dr. Cullen and other physicians began to throw off the Boerhaavian yoke; but from the minute observation which Dr. Darwin has given of the laws of association, habits and phenomena of animal life, it is manifest that his system is the result of the operation of his own mind.

The only other record of his life in Edinburgh which I possess is a letter to his friend Dr. Okes, of Exeter,[4] which shows his sceptical frame of mind when twenty-three years old. The expression "disagreeable news" applied to his father's death, sounds very odd to our ears, but he evidently used this word where we should say "painful". For, in a feeling letter to Josiah Wedgwood, written a quarter of a century afterwards (Nov. 29, 1780) about the death of their common friend Bentley, in which he alludes to the death of his own son, he says nothing but exertion will dispossess "the *disagreeable* ideas of our loss".

ERASMUS DARWIN TO DR. OKES

Yesterday's post brought me the disagreeable news of my father's departure out of this sinful world. {Here follows the character of his father already quoted.}

He was 72 years old, and died the 20th of this current November 1754. 'Blessed are they that die in the Lord'.

That there exists a superior ENS ENTIUM, which formed these wonderful creatures, is a mathematical demonstration. That HE influences things by a particular providence, is not so evident. The probability, according to my notion, is against it, since general laws seem sufficient for that end. Shall we say no particular providence is necessary to roll this Planet round the Sun, and yet affirm it necessary in turning up *cinque* and *quatorze*, while shaking a box of dies? or giving each his daily bread? The light of Nature affords us not a single argument for a future state; this is the only one, that it is possible with God, since He who made us out of nothing can surely re-create us; and that He will do this is what we humbly hope. I like the Duke of Buckingham's epitaph— "Pro Rege sæpe, pro Republicâ semper, dubius, non improbus vixi; incertus, sed inturbatus morior. Christum advenero, Deo confido benevolenti et omnipotenti, Ens Entium miserere mei!"

Erasmus Darwin

In 1755 he returned to Cambridge, and took his Bachelor of Medicine degree. He then returned to Edinburgh, and early in September 1756, settled as a physician in Nottingham. Here, however, he remained for only two or three months, as he got no patients. Whilst in Nottingham he wrote several letters, some in Latin and some in English, to his friend, the son of the famous German philosopher, Reimarus.[5] Mechanics and medicine were the bonds of union between them. Erasmus also dedicated a poem to young Reimarus, on his taking his degree at Leyden in 1754. Various subjects were discussed between them, including the wildest speculations by Erasmus on the resemblance between the action of the human soul and that of electricity, but the letters are not worth publishing. In one of them he says:

> I believe I forgot to tell how Dr. Hill makes his 'Herbal' {a formerly
> well-known book}. He has got some wooden plates from some
> old herbal, and the man that cleans them cuts out one branch of
> every one of them, or adds one branch or leaf, to disguise them.
> This I have from my friend Mr. G—y [Gurney], watch-maker, to
> whom this print-mender told it, adding, 'I make plants now every
> day that God never dreamt of'.

It also appears from one of his letters to Reimarus that Erasmus corresponded at this time about short-hand writing with Gurney, the author of a famous book on this art, which my grandfather practised for some time, so that whilst young he filled six volumes with short-hand notes.

Several of the letters to Reimarus relate to a case which is partly to the credit and partly to the discredit of my grandfather. He seems to have been much interested about a working man whom he sent or helped to send to a London surgeon, Mr. D., for a serious operation. There appears to have been some misunderstanding between Dr. Darwin, Reimarus, and the surgeon, who they expected would perform the operation gratuitously. Dr. Darwin wrote to Reimarus:

> I am very sorry to hear that D. took six guineas from the poor
> young man. He has nothing but what hard labour gives him; is
> much distressed by this thing costing him near £30 in all, since the

house where he lay cheated him much. . . .When he returns I shall send him two guineas. I beg you would not mention to my brother that I send this to him.

Why his brother should not be told of this act of charity it is difficult to conjecture.

From two other letters it appears that my grandfather wrote an anonymous letter to the surgeon, complaining of his charge; and this, though in a good cause, was a discreditable action, which he ought to have fully owned when suspected of being the author. Erasmus, on hearing that he was suspected, wrote to Reimarus (Nottingham, Sept. 9, 1756), adding in a P.S. "that he might show the letter to Mr. D."

> You say I am suspected to be the Author of it {i.e. the anonymous letter}, and next to me some malicious Person somewhere else, and that I am desired as I am a gentleman to declare concerning it. First, then, as I am upon Honour, I must not conceal that I am glad there are Persons who will revenge Faults, that Law can not take hold off: and I hope Mr. D. will not be affronted at this Declaration; since you say he did not know the Distress of the Man. Secondly, as another Person is suspected, I will not say whether I am the Author or not, since I don't think the Author merits Punishment, for informing Mr. D. of a Mistake. You call the Letter a threatening Letter, and afterwards say the Author pretends to be a Friend to Mr. D. This, though you give me several particulars of it, is a Contradiction I don't understand.

The anonymous letter answered its purpose, for the surgeon returned four guineas, and my grandfather thought it probable that he would ultimately return the other two guineas.

In November 1756, Erasmus settled in Lichfield, and now his life may be said to have begun in earnest; for it was here, and in or near Derby, to which place he removed in 1781, that he published all his works.

Owing to two or three very successful cases, he soon got into some practice at Lichfield as a physician, when twenty-five years old.

A year afterwards (December 1757) he married Miss Mary Howard, aged 17–18 years, who, judging from all that I have heard of her, and from some of her letters, must have been a superior and charming woman. She died after a long and suffering illness in 1770. They seem to have lived together most happily during the thirteen years of their married life, and she was tenderly nursed by her husband during her last illness. Miss Seward gives,[6] on second-hand authority, a long speech of hers, ending with the words, "he has prolonged my days, and he has blessed them". This is probably true, but everything which Miss Seward says must be received with caution; and it is scarcely possible that a speech of such length could have been reported with any accuracy.

The following letter was written by Erasmus four days before his marriage with Miss Howard, and I give it because any one who judged of his character from some of the statements made by Miss Seward, might well doubt whether he possessed any human feelings.

ERASMUS DARWIN TO MARY HOWARD

Darlaston, Dec. 24, 1757.

Dear Polly,

As I was turning over some old mouldy volumes, that were laid upon a Shelf in a Closet of my Bed-chamber, one I found, after blowing the Dust from it with a Pair of Bellows, to be a Receipt Book, formerly, no doubt, belonging to some good old Lady of the Family. The Title Page (so much of it as the Rats had left) told us it was "a Bouk off verry monny muckle vallyed Receipts bouth in Kookery and Physicks". Upon one Page was "To make Pye-Crust",—in another "To make Wall-Crust",—"To make Tarts",—and at length "To make Love". "This Receipt", says I, "must be curious, I'll send it to Miss Howard next

Post, let the way of making it be what it will".—Thus it is. "To make Love. Take of Sweet-William and of Rose-Mary, of each as much as is sufficient. To the former of these add of Honesty and Herb-of-grace; and to the latter of Eye-bright and Motherwort of each a large handful: mix them separately, and then, chopping them altogether, add one Plumb, two sprigs of Heart's Ease and a little Tyme. And it makes a most excellent dish, probatum est. Some put in Rue, and Cuckold-Pint, and Heart-Chokes, and Coxcome, and Violents; But these spoil the flavour of it entirely, and I even disprove of Sallery which some good Cooks order to be mix'd with it. I have frequently seen it toss'd up with all these at the Tables of the Great, where no Body would eat of it, the very appearance was so disagreable."

Then follow'd "Another Receipt to make Love", which began "Take two Sheep's Hearts, pierce them many times through with a Scewer to make them Tender, lay them upon a quick Fire, and then taking one Handful—" here Time with his long Teeth had gnattered away the remainder of this Leaf. At the Top of the next Page, begins "To make an honest Man". "This is no new dish to me", says I, "besides it is now quite old Fashioned; I won't read it". Then follow'd "To make a good Wife". "Pshaw", continued I, "an acquaintance of mine, a young Lady of Lichfield, knows how to make this Dish better than any other Person in the World, and she has promised to treat me with it sometime", and thus in a Pett threw down the Book, and would not read any more at that Time. If I should open it again tomorrow, whatever curious and useful receipts I shall meet with, my dear Polly may expect an account of them in another Letter.

I have the Pleasure of your last Letter, am glad to hear thy cold is gone, but do not see why it should keep you from the concert, because it was gone. We drink your Health every day here, by the Name of Dulcinea del Toboso, and I told Mrs. Jervis and Miss Jervis that we were to have been married yesterday, about which they teased me all the Evening. I heard nothing of Miss Fletcher's Fever before.

I will certainly be with Thee on Wednesday evening, the Writings are at my House, and may be dispatched that night, and if a License takes up any Time (for I know nothing at all about these Things) I should be glad if Mr. Howard would order one, and by this means, dear Polly, we may have the Ceremony over next morning at eight o'clock, before any Body in Lichfield can know almost of my being come Home. If a License is to be had the Day before, I could wish it may be put off till late in the Evening, as the Voice of Fame makes such quick Dispatch with any News in so small a Place as Lichfield.—I think this is much the best scheme, for to stay a few Days after my Return could serve no Purpose, it would only make us more watch'd and teazed by the Eye and Tongue of Impertinence.

I shall by this Post apprize my Sister to be ready, and have the House clean, and I wish you would give her Instructions about any trivial affairs, that I cannot recollect, such as a cake you mentioned, and tell her the Person of whom, and the Time when it must be made, &c. I'll desire her to wait upon you for this Purpose. Perhaps Miss Nelly White need not know the precise Time till the Night before, but this as you please, as I [expect(?)] you could rely upon her Secrecy, and it's a Trifle, if any Body should know.

Matrimony, my dear Girl, is undoubtedly a serious affair (if any Thing be such), because it is an affair for Life. But, as we have deliberately determin'd, do not let us be *frighted* about this Change of Life; or however, not let any breathing Creature perceive that we have either Fears or Pleasures upon this Occasion: as I am certainly convinced, that the best of Confidants (tho' experienced on a thousand other Occasions) could as easily hold a burning cinder in their Mouth as anything the least ridiculous about a new married couple!

I have ordered the Writings to be sent to Mr. Howard that he may peruse and fill up the blanks at his Leizure, as it will (I foresee) be dark night before I get to Lichfield on Wednesday.

Mrs. Jervis and Miss desire their Compl. to you, and often say how glad she shall be to see you for a few Days at any Time.

I shall be glad, Polly, if thou hast Time on Sunday night, if thou wilt favour me with a few Lines by the return of the Post, to tell me how Thou doest, &c.—My Compl. wait on Mr. Howard if He be returned.—My Sister will wait upon you, and I hope, Polly, Thou wilt make no Scruple of giving her Orders about whatever you chuse, or think necessary. I told her Nelly White is to be Bride-Maid. Happiness attend Thee! adieu,

from, my dear Girl,

thy sincere Friend,

E. Darwin.

P.S.—Nothing about death in this Letter, Polly.

It has been said that he soon got into practice at Lichfield, and I have found the following memorandum of his profits in his own handwriting:

The profits of my business amounted

		£	s.	d.
From Nov. 12, 1756 to Jan. 1, 1757		18	7	6
Jan. 1757	Jan. 1758	192	10	6
" 1758	" 1759	305	2	0
" 1759	" 1760	469	4	0
" 1760	" 1761	544	2	0
" 1761	" 1762	669	18	0
" 1762	" 1763	726	7	0
" 1763	" 1764	639	13	0
" 1764	" 1765	750	13	0
" 1765	" 1766	800	1	4
" 1766	" 1767	748	5	6
" 1767	" 1768	847	3	0
" 1768	" 1769	775	11	6
" 1769	" 1770	?		
" 1770	" 1771	956	17	6
" 1771	" 1772	1064	7	6
" 1772	" 1773	1025	3	0

Later in life he gave up the good habit of keeping accurate accounts, for in 1799 he wrote to my father that he had been much perplexed what return to make to the commissioners {of income tax?}, as "I kept no book, but believed my business to be £1000 a year, and deduct £200 for travelling expenses and chaise hire, and £200 for a livery-servant, four horses and a day labourer". Subsequently he informed my father that the commissioners had accepted this estimate. A century ago an income of £1000 would probably be equal to one of £2000 at the present time; but I am surprised that his profits were not larger. *There is, however, an interval of twenty-six years between the beginning of 1773 and 1799, by which time, with advancing years and several changes of residence, his practice had probably decreased. In this interval his profits must have been greater than those in the figures given, judging from what he was able to bequeath to his children by his second marriage.* All his friends constantly refer to his long and frequent journeys, for his practice lay chiefly amongst the upper classes of society. When he went to live at the Priory, he remarked to my father in a letter that five or six additional miles would make little difference in the fatigue of his journeys.

After settling at Lichfield, he attended, during several years, chiefly to medicine; but no doubt he was continually observing and making notes on various subjects. A huge folio common-place book, begun in 1776, is in the possession of Reginald Darwin, and is half filled with notes and speculations.

Considering how voluminous a writer he became when old, it is remarkable that he does not appear to have thought for a long time of publishing either prose or poetry. In a letter dated Nov. 21, 1775 to Mr. Cradock,[7] thanking him for a present of his 'Village Memoirs', he says:

> I have for twenty years neglected the muses, and cultivated medicine alone with all my industry . . . I lately interceded with a Derbyshire lady to desist from lopping a grove of trees, which has occasioned me . . . to try again the long-neglected art of verse-making, which I shall inclose to amuse you, promising, at the same time, never to write another verse as long as I live, but to apply my time to finishing a work on some branches of medicine, which I intend for a posthumous publication.

It may be convenient here to mention that in 1781, eleven years after the death of his first wife, Dr. Darwin married the widow of Colonel Edward Pole [C.D. has 'Colonel Chandos Pole'] of Radburn Hall. He had become acquainted with her in the spring of 1778, when she had come to Lichfield in order that he might attend her children professionally. On his second marriage he left Lichfield and, after living two years at Radburn Hall, he removed into the town of Derby, and ultimately to Breadsall Priory, a few miles from the town, where he died in 1802.

There is little to relate of his life either at Lichfield or Derby, except his works.

In 1778 he purchased the lease of a pretty valley, about eight acres in extent, near Lichfield, and made it into a botanic garden; and this seems to have been his chief amusement. Miss Seward describes[8] the place in her grandiose style as "a wild umbrageous valley . . . irriguous from various springs, and swampy from their plenitude". It now forms part of an adjoining park; and a Handbook for Lichfield says it is still "a wild spot, but very picturesque; many of the old trees remaining, and occasionally a few Darwinian snow-drops and daffodils peeping through the turf, and bravely fighting the battle of life".

This garden led him to write his poem of the 'Botanic Garden', the second part of which, entitled the 'Loves of the Plants', was published, oddly enough, before the first part, called the 'Economy of Vegetation'. The 'Loves of the Plants', judging from a prefixed sonnet, must have appeared in 1788, and the second edition in 1790. In a letter to my father, dated Feb. 21, 1788, he says: "I am printing the Loves of the Plants, which I shall not put my name to, tho' it will be known to many. But the addition of my name would seem as if I thought it a work of consequence."

Notwithstanding this depreciatory estimate, its success was great and immediate; and I have heard my father, who was accurate about figures, say that a thousand guineas were paid before publication for the part which appeared as the second; and such a sum must have been something extraordinary in those days. Nor was the success quite transitory, for a fourth edition appeared in 1799. In 1806 an octavo edition of all his poetical works was published in three volumes.

I have myself met with old men who spoke with a degree of enthusiasm about his poetry, quite incomprehensible at the present day. Horace Walpole, a highly-cultivated man and a master of the English language, in his letters repeatedly alludes with admiration to Dr. Darwin's poetry, and in a letter to Mr. Barrett (May 14, 1792) writes:

> The 'Triumph of Flora', beginning at the fifty-ninth line, is most beautifully and enchantingly imagined; and the twelve verses that by miracle describe and comprehend the creation of the universe out of chaos, are in my opinion the most sublime passages in any author, or in any of the few languages with which I am acquainted. There are a thousand other verses most charming, or indeed all are so, crowded with most poetic imagery, gorgeous epithets and style: and yet these four cantos do not please me equally with the 'Loves of the Plants'.

The lines thus eulogised are:[9]

> 'Let there be light!' proclaim'd the Almighty Lord.
> Astonish'd Chaos heard the potent word.
> Through all his realms the kindling Ether runs,
> And the mass starts into a million suns;
> Earths round each sun with quick explosions burst,
> And second planets issue from the first;
> Bend, as they journey with projectile force,
> In bright ellipses their reluctant course;
> Orbs wheel in orbs, round centres centres roll,
> And form, self-balanced, one revolving whole.
> Onward they move amid their bright abode,
> Space without bound, the bosom of their God!

Another highly-cultivated yet very different man, R. L. Edgeworth, in a letter (1790) to Dr. Darwin, writes thus about the 'Botanic Garden':[10]

> I may, however, without wounding your delicacy, say that it has silenced for ever the complaints of poets, who lament that Homer, Milton, Shakespeare, and a few classics, had left nothing new to describe, and that elegant imitation of imitations was all that could be expected in modern poetry . . . I read the description of the Ballet of Medea to my sisters, and to eight or ten of my own family. It seized such hold of my imagination, that my blood thrilled back through my veins, and my hair broke the cementing of the friseur, to gain the attitude of horror.

After the fame of his poetry had begun to wane, Edgeworth predicted (p.117) "that in future times some critic will arise who shall rediscover the 'Botanic Garden', and build his fame upon the discovery. . . . It will shine out again, the admiration of posterity."

Several poets addressed odes in his honour, as may be seen in the edition of 1806. Cowper, who would now be ranked by everyone as incomparably above Darwin as a poet, and who, one would have thought, differed in taste from him as much as two men could differ, yet, in conjunction with Hayley, wrote a poem in his honour[11] of which I will give one stanza:

> No envy mingles with our praise,
> Tho' could our hearts repine
> At any poet's happier lays,
> They would, they must, at thine.

In the second part of this volume we may read the judgment at the present time of the poetry of Dr. Darwin by a highly-cultivated foreigner.

Notwithstanding the former high estimation of his poetry by men of all kinds in England, no one of the present generation reads, as it appears, a single line of it. So complete a reversal of judgment within a few years is a remarkable phenomenon. His verses were, however, quizzed by some persons not long after their publication. In the 'Pursuits of Literature',[12] they are called:

Filmy, gauzy, gossamery lines. . . .
Sweet tetrandrian, monogynian strains.

But the downfall of his fame as a poet was chiefly caused by the publication of the well-known parody the 'Loves of the Triangles'. No doubt public taste was at this time changing, and becoming more simple and natural. It was generally acknowledged, under the guidance of Wordsworth and Coleridge, that poetry was chiefly concerned with the feelings and deeper workings of the mind; whereas, Darwin maintained that poetry ought chiefly to confine itself to the word-painting of visible objects. He remarks[13] that poetry should consist of words which express ideas originally received by the organ of sight.

. . . And as our ideas derived from visible objects are more distinct than those derived from the objects of our other senses, the words expressive of these ideas belonging to vision make up the principal part of poetic language. That is, the poet writes principally for the eye; the prose writer uses more abstracted terms.

That Darwin was a great master of language will hardly be denied. In one of the earliest and best criticisms on his poetry[14] it is said no man "had a more imperial command of words, or could elucidate with such accuracy and elegance the most complex and intricate machinery". Even Byron called him "a mighty master of unmeaning rhyme".

His first scientific publication was a paper in the 'Philosophical Transactions' for 1757, in which he confutes the view of Mr. Eeles, that vapour ascends through "every particle being endued with a portion of electric fire". The paper is of no value, but is curious as showing in what a rudimentary condition some branches of science then were. For Dr. Darwin remarks that the "distinction has not been sufficiently considered by anyone to my knowledge" between "the immense rarefaction of explosive bodies" due "to the escape of air before condensed in them", as when a few grains of gunpowder are exploded in a bladder, and to "the expansion of the constituent parts of those bodies" through heat, as with steam.

The 'Zoonomia', which had been in preparation during many years, was published in 1794. We have seen that in 1775 it was intended for posthumous publication. Even so late as February 1792, my grandfather wrote to my father: "I am studying my 'Zoonomia', which I think I shall publish, in hopes of selling it, as I am now too old and hardened to fear a little abuse. Every John Hunter must expect a Jessy Foot to pursue him, as a fly bites a horse." The work when published was translated into German, French, and Italian, and was honoured by the Pope by being placed in the 'Index Expurgatorius'.

Dr. Krause has given so full, impartial, and interesting an account of the scientific views contained in this and his other works that I need say little on this head.

Although Dr. Darwin indulged largely in hypotheses, he knew full well the value of experiments. Maria Edgeworth, in writing (March 9, 1792) about her little brother Henry, who was fond of collecting and observing, says: "He will at least never come under Dr. Darwin's definition of a fool. A fool, Mr Edgeworth, you know, is a man who never tried an experiment in his life".[15]

Again, in an 'Apology' prefixed to the 'Botanic Garden', we have the following just remarks:

It may be proper here to apologise for many of the subsequent conjectures on some articles of natural philosophy, as not being supported by accurate investigation, or conclusive experiments. Extravagant theories, however, in those parts of philosophy where our knowledge is yet imperfect, are not without their use; as they encourage the execution of laborious experiments, or the investigation of ingenious deductions to confirm or refute them. And since natural objects are allied to each other by many affinities, every kind of theoretic distribution of them adds to our knowledge by developing some of their analogies.

Dr. Darwin proved himself more ready to admit those new and grand views in chemistry (a branch of science which always greatly interested him) which were developed towards the close of the last century, than some professed chemists. James Keir, a distinguished chemist of the day, writing to him on March 15, 1790, says:[16]

> You are such an infidel in religion that you cannot believe in transubstantiation, yet you can believe that apples and pears, &c., sugar, oil, vinegar, are nothing but water and charcoal, and that it is a great improvement in language to call all these things by one word—oxyde hydro-carbonneux.

There is a good deal of psychology in the 'Zoonomia', but I fear that his speculations on this subject cannot be ranked as of much value. Nevertheless, G. H. Lewes says:[17] "Although even more neglected than Hartley by the present generation, Darwin, once so celebrated, deserves mention here as one of the psychologists who aimed at establishing the physiological basis of mental phenomena". Although Lewes afterwards adds:

> Had Darwin left us only the passage just cited[18] we should have credited him with a profounder insight into psychology than any of his contemporaries and the majority of his successors exhibit; and although the perusal of 'Zoonomia' must convince everyone that Darwin's system is built up of absurd hypotheses, Darwin deserves a place in history for that one admirable conception of psychology as subordinate to the laws of life.

It may be added that the illustrious Johannes Müller quotes with approbation, though with correction, Darwin's 'Law of Associated Movements'.[19]

The 'Zoonomia' is largely devoted to medicine, and my father thought that it had much influenced medical practice in England; he was of course a partial, yet naturally a more observant judge than others on this point. The book when published was extensively read by the medical men of the day, and the author was highly esteemed by them as a practitioner.

The following curious story, written down by his daughter Violetta in her old age, shows his repute as a physician. A gentleman in the last stage of consumption came to Dr. Darwin at Derby, and expressed himself to this effect:

> I am come from London to consult you, as the greatest physician in the world, to hear from you if there is any hope in my case; I know that my life hangs upon a thread, but while there is life there may be hope. It is of the utmost importance for me to settle my worldly affairs immediately; therefore I trust that you will not deceive me, but tell me without hesitation your candid opinion.

Dr. Darwin felt his pulse, and minutely examined him, and said he was sorry to say there was no hope. After a pause of a few minutes the gentleman said: "How long can I live?" The answer was: "Perhaps a fortnight". The gentleman seized Dr. Darwin's hand and said: "Thank you, doctor, I thank you; my mind is satisfied; I now know there is no hope for me". Dr. Darwin then said: "But as you come from London, why did you not consult Dr. Warren, so celebrated a physician?" "Alas! doctor, I am Dr. Warren." He died in a week or two afterwards.

I will give another small incident showing how highly the memory of Dr. Darwin was esteemed for more than twenty years after his death by medical men. One of his granddaughters presented a case of some perplexity, and was taken by her mother separately to three of the most eminent London surgeons, whose names are still well-known; and the mother, in order to excite their interest, mentioned that she herself was the daughter of Dr. Darwin. Whereupon everyone of these surgeons declined at first to take his fee; and they all expressed the pleasure they felt in seeing the daughter of one to whose works they owed so much.

I remember only two points, with respect to which my father thought that medical practice in this country had been influenced by the 'Zoonomia'.[20] In this work it is said:

> There is a golden rule by which the necessary and useful quantity of stimulus in fevers with debility may be ascertained. When wine or beer is exhibited, either alone or diluted with water, if the pulse becomes slower the stimulus is of a proper quantity, and should be repeated every two or three hours, or when the pulse again becomes quicker.

The value of this golden rule will be appreciated when it is remembered how highly stimulants are now valued in fever. His views on fever certainly attracted attention at the time;[21] but the use of stimulants in such cases has fluctuated much, and the history of the subject is an obscure one, as I infer from a letter which Sir Robert Christison has had the kindness to send me.

The second point mentioned by my father, was the treatment of the insane, and here we are concerned with humanity as well as with medical practice. After saying[22] that no lunatic should be restrained unless he be dangerous, Dr. Darwin urges that with other insane patients "confinement retards rather than promotes their cure, which is forwarded by change of ideas, &c." He then remarks that mistaken ideas do not by themselves justify confinement, and adds: "If everyone who possesses mistaken ideas, or who puts false estimates on things, was liable to confinement, I know not who of my readers might not tremble at the sight of a madhouse".

In connection with this subject the following quotation from Dr. Maudsley is interesting:[23]

> Here I may fitly take occasion to adduce certain observations with regard to the striking manner in which diseased action of one nervous centre is sometimes transferred suddenly to another, a fact which, though it has lately attracted new attention, was long since noticed and commented on by Dr. Darwin: "In some convulsive diseases", he writes, "a delirium or insanity supervenes, and the

convulsions cease; and conversely, the convulsions shall
supervene, and the delirium cease. Of this I have been a witness
many times a day, in the paroxysms of violent epilepsies, which
evinces that one kind of delirium is a convulsion of the organs of
sense; and that our ideas are the motions of these organs."

Dr. Lauder Brunton has mentioned to me another instance
in which Dr. Darwin apparently anticipated a modern discovery.
In an article in the 'British Medical Journal' (1873, p. 735) on
"catching cold", an account is given of Rosenthal's experiments,
showing that when an animal is exposed to a rather high temperature, "the cutaneous vessels become paralysed by the heat, and
remain dilated even after cold has been applied. The blood is thus
exposed over a large surface, and becomes rapidly cooled." For
instance, the blood of an animal thus treated fell from between
$107\cdot6°$ and $111°$ to $96\cdot8°$, and remained at this lower temperature
for several days. A passage in the 'Zoonomia'[24] seems to show that
Dr. Darwin was acquainted with the above important fact, discovered by Rosenthal in 1872, and this is a remarkable instance of
his acute powers of observation.

Finally, Dr. Darwin fully recognized the truth and importance of
the principle of inheritance in disease. He remarks;[25] "As many
families become gradually extinct by hereditary diseases, as by
scrofula, consumption, epilepsy, mania, it is often hazardous to
marry an heiress, as she is not unfrequently the last of a dis-
eased family". This is a remark which his grandson, Francis Galton, would fully appreciate. On the other hand, when a tendency
to disease is confined to one parent, the children often escape.
"I now know", as Dr. Darwin writes to my father, January 5th, 1792,
"many families who had insanity on *one* side, and the children,
now old people, have had no symptom of it. If it were otherwise,
there would not be a family in the kingdom without epileptic,
gouty, or insane people in it."

In 'The Temple of Nature' (Notes, p.11), there is a curious
instance of his prophetic sagacity with respect to 'microscopic

animals'. A few years since, a *cui bono* philosopher might have sneered at men spending their lives in the examination of organisms far too minute to be seen by the naked eye; and it would have been difficult to have given such a man any satisfactory answer except on general principles. But we now know from the researches of various naturalists how all-important a part these organisms play in putrefaction, fermentation, infectious diseases, &c.; and as a consequence of such researches, the world owes a deep debt of gratitude to Mr. Lister for his anti-septic treatment of wounds. Therefore the following sentence of my grandfather, considering how little was then known on the subject, appears to me remarkable. He says: "I hope that microscopic researches may again excite the attention of philosophers, as unforeseen advantages may probably be derived from them like the discovery of a new world".

As Dr. Krause has not said much about the 'Phytologia', a few remarks may be made on this work, published in 1800. It begins with a discussion on the nature of leaf-buds and flower-buds; and the view, now universally adopted, that a plant consists of "a system of individuals", and not merely of a multiplication of similar organs, originated with Darwin, as I infer from Johannes Müller's 'Elements of Physiology'.[26]

Considering how recently the manner in which plants modify and absorb the nutriment stored up in their roots, tubers, cotyledons, &c., has been understood, the following sentence ('Phytologia', p. 77) deserves notice:

> The digestive powers of the young vegetable, with the chemical agents of heat and moisture, convert the starch or mucilage of the root or seed into sugar for its own nourishment; . . . and thus it appears probable that sugar is the principal nourishment of both animal and vegetable beings.

The work treats largely of agriculture and horticulture, and a section is devoted to phosphorus, which, as he believes (p. 207), exists universally in vegetables, a subject "which has not yet been sufficiently attended to". He then refers to the use of bones as a manure, but erred in supposing that shells and some other substances which are luminous in the dark, abounded with phosphorus. Sir J. Sinclair, President of the Board of Agriculture, and therefore a most capable judge, says that though the fertilising properties of bone-dust had been previously noticed by Hunter, yet "they were first theoretically explained and brought forward with authority by Dr. Darwin". He then remarks, and of the truth of this remark there can be little doubt, "perhaps no {other} modern discovery has contributed so powerfully to improve the fertility and to increase the produce of the soil".[27]

The following sentences are interesting as forecasting the progress of modern thought. In a discussion on 'The Happiness of Organic Life' ('Phytologia', p. 556), after remarking that animals devour vegetables, Dr. Darwin says:

> The stronger locomotive animals devour the weaker ones without mercy. Such is the condition of organic nature! whose first law might be expressed in the words, 'eat or be eaten', and which would seem to be one great slaughterhouse, one universal scene of rapacity and injustice.

He proceeds: "Where shall we find a benevolent idea to console us amid so much apparent misery?" He then argues:

> Beasts of prey more easily catch and conquer the aged and infirm, and the young ones are defended by their parents. . . . By this contrivance more pleasurable sensation exists in the world . . . old organisations are transmigrated into young ones . . . death cannot so properly be called positive evil as the termination of good.

There is much more of the same kind, and hardly more relevant. He then makes a great leap in his argument, and concludes that all the strata of the world "are monuments of the past felicity of organised nature! and consequently of the benevolence of the deity!"

Finally, it is a curious proof how English botanists had been blinded by the splendour of the fame of Linnæus, that Dr. Darwin apparently had never heard of Jussieu, for he writes (p. 564):

> If the system of the great Linnæus can ever be intrinsically improved, I am persuaded that the plan here proposed of using the situations, proportions, or forms, with or without the number of the sexual organs, as criterions of the orders and classes, must lay the foundation; but that it must require a great architect to erect the superstructure.

He therefore did not know that a noble superstructure had already been raised.

There remains only one other book to notice: 'A Plan for the Conduct of Female Education in Boarding Schools', published in 1797. This treatise seems never to have received much attention in England, though it was translated into German. It is strongly characterised throughout by plain common sense, with little theorising, and is everywhere benevolent. He insists that punishment should be avoided as much as possible, and that reproof should be given with kindness. Emulation, though useful, is dangerous, from being liable to degenerate into envy.

> If once you can communicate to children a love of credit and an apprehension of shame, you have instilled into them a principle, which will constantly act and incline them to do right, though it is not the true source whence our actions ought to spring, which should be from our duty to others and ourselves.

He urges that sympathy with the pains and pleasures of others is the foundation of all our social virtues; and that this can best be inculcated by example and the expression of our own sympathy. "Compassion, or sympathy with the pains of others, ought also to extend to the brute creation . . . to destroy even insects wantonly shows an unreflecting mind, or a depraved heart".

He considers it of great importance to girls that they should learn to judge of character, as they will some day have to choose a husband; and he believes that the reading of proper novels teaches them something of life and mankind, and helps them to avoid mistakes in judging of character. He also remarks more than once, that children express various emotions in their countenances much more plainly than do older persons; and he is convinced that one great advantage which a child derives from going to school is in unconsciously acquiring a knowledge of physiognomy through mixing with other children. This knowledge, "by giving a promptitude of understanding the present approbation or dislike, and the good or bad designs of those whom we converse with, becomes of hourly use in almost every department of life".

Remembering when this book was published, namely in the last century, we see that he was much in advance of his age in his ideas as to sanitary arrangements—such as supplying towns with pure water, having holes made into crowded sitting and bed-rooms for the constant admission of fresh air, and not allowing chimneys to be closed during summer—and as to diet and exercise. He speaks of "skating on the ice in winter, swimming in summer, funambulation or dancing on the straight rope", as "not allowed to ladies by the fashion of this age and country". It is a pity he does not tell us when and where it was the fashion of young ladies to funambulate! With respect to swimming, he disregarded fashion, and his own daughters and sons were taught to swim at a very early age, so that they became, it is said, expert in this art when only four years old.

In the 'Phytologia' he shows himself still more clearly a great sanitary reformer. He insists that the sewage from towns, which are now left buried or carried into the rivers, should be removed for the purpose of agriculture; "and thus the purity and healthiness of the towns may contribute to the thriftiness and wealth of the surrounding country". Also, "there should be no burial places in churches or in churchyards, where the monuments of departed sinners shoulder God's altar, . . . but proper burial grounds should be consecrated out of towns". Nearly a century has elapsed since this good advice was given, and it has as yet only partially been carried out.

———————————— ⌒ ————————————

One of the subjects which interested Dr. Darwin most throughout his whole life, and which appears little in his published works, was mechanical invention. This is shown in his letters to Josiah Wedgwood, Edgeworth and others, and in a huge common-place book full of sketches and suggestions about machines. He seems, however, rarely to have completed anything, with the exception of a horizontal windmill for grinding flints, which he designed for Wedgwood, and which answered its purpose.

There are schemes and sketches for an improved lamp, like our present moderators; candlesticks with telescope stands so as to be raised at pleasure to any required height; a manifold writer; a knitting loom for stockings; a weighing machine; a surveying machine; a flying bird, with an ingenious escapement for the movement of the wings, and he suggests gunpowder or compressed air as the motive power.

He also gives a plan of a canal lock, on the principle of the boat being floated into a large box, the door of which is then closed, and the box afterwards raised or lowered. This principle has since been acted on under certain circumstances, but by an improved method.

A rotatory pump was also one of his schemes, and this, likewise, under a modified form, is extensively used for blowing air into cupolas, and for pumping water in certain cases.

He saw clearly, as he explains in 1756 in a letter to Reimarus, that it would be a great advantage if the spokes of carriage wheels acted as springs; and Sir. J. Whitworth has recently had a carriage constructed with such wheels, which is remarkably smooth.

Another invention was a small carriage of peculiar construction, intended to give the best effect to the power of the horse, combined with the greatest ease in turning. According to Miss Seward,

It was a platform, with a seat fixed upon a very high pair of wheels, and supported in the front upon the back of the horse, by means of a kind of proboscis, which, forming an arch, reached over the hind quarters of the horse; and passed through a ring, placed on an upright piece of iron, which worked in a socket, fixed in the saddle.

{Dr. Krause informs me "that the Moravian engineer, Theodo Tomatschek, has lately constructed a very similar carriage, which I saw at the Vienna International Exhibition; and the Americans have also reduced the Darwinian idea to practice, and given the new vehicle the paradoxical name 'Equibus'".} But, however correct this carriage may have been in principle, Darwin had the misfortune, in the year 1768, to be upset in it, when he broke his knee-cap and ever afterwards limped a little.

A speaking machine was a favourite subject, and for this end he invented a phonetic alphabet. His machine, or "head, pronounced the p, b, m, and the vowel a, with so great nicety as to deceive all who heard it unseen, when it pronounced the words mama, papa, map, and pam; and it had a most plaintive tone, when the lips were gradually closed".[28] Edgeworth also bears witness to the capacity of this speaking head. Matthew Boulton entered into the following agreement, which, from the witnesses to it, was evidently made at one of the meetings of the famous Lunar Club; but whether in joke or earnest, it is difficult to conjecture:

> I promise to pay to Dr. Darwin of Lichfield one thousand pounds upon his delivering to me (within 2 years from date hereof) an Instrument called an organ that is capable of pronouncing the Lord's Prayer, the Creed, and Ten Commandments in the Vulgar Tongue, and his ceding to me, and me only, the property of the said invention with all the advantages thereunto appertaining.
>
>
>
> M. Boulton
> Soho Sep. 3rd 1771
> Witness James Keir
> Witness W. Small
>
>

In the last century a speaking tube was an unknown invention in country districts, and my grandfather had one for his study, which opened near the back of the kitchen fireplace. A countryman had brought a letter and sat by this fire, which had become very low, whilst waiting for an answer, when suddenly he heard a sepulchral voice, saying "I want some coals". The man instantly fled from the house, for my grandfather had the reputation amongst the country folk of being a sort of magician.

At a time (1783) when very few artesian wells had been made in this country, my grandfather made one, though on a small scale; and in the garden-wall to his house in Full Street, Derby, there still exists an iron plate with the following inscription:

TEREBELLO EDUXIT AQUAM
ANNO MDCCLXXXIII.
ERASMUS DARWIN.
LABITUR ET LABETUR.

This case would not have been worth mentioning had he not shown in his paper,[29] in which this well is described, that he recognised the true principles of artesian wells. He remarks that "some of the more interior strata of the earth are exposed naked on the tops of mountains; and that in general, those strata which lie uppermost, or nearest to the summit of the mountain, are the lowest in the contiguous plains". He then adds that the waters "sliding between two of the strata above described, descend till they find or make for themselves an outlet, and will in consequence rise to a level with the parts of the mountain where they originated".

In October 1771 he wrote several letters to Wedgwood about a scheme of making, with his own capital, a canal of very small dimensions from the Grand Trunk to Lichfield, for boats drawing only a foot of water, to be dragged by a man, and carrying only four or five tons burthen. Such a canal would have borne the same relation to ordinary canals, as some very narrow railways, which have been found to answer well in Wales, bear to ordinary railways. He seems to have been greatly interested in this project, which, however, never came to anything.

The weather, and the course of the winds throughout the world, was another subject on which he was continually searching for information and speculating. I have heard my father say that, in order to notice every change of the wind, he connected a wind-vane on the top of his house with a dial on the ceiling of his study.

Here will be a convenient place to give a very few of his letters, or extracts from them, as they may serve to illustrate some points in his character. His correspondence with many distinguished men was large; but although I possess many of his letters, and have seen others, they are mostly uninteresting, and not worth publication. Medicine and mechanics alone roused him to write with any interest.

He occasionally corresponded with Rousseau, but none of their letters have been preserved. He became acquainted with him in an odd manner. Rousseau was living in 1766 at Mr. Davenport's house, Wootton Hall, and he used to spend much of his time "in the well-known cave upon the terrace in melancholy contemplation". He disliked being interrupted, so my grandfather, who was then a stranger to him, sauntered by the cave, and minutely examined a plant growing in front of it. This drew forth Rousseau, who was interested in botany, and they conversed together, and afterwards corresponded during several years.

In February 1767 a gentleman consulted my grandfather about a baby, apparently the illegitimate child of a lady, which, it was suspected, had been murdered by its mother; *and my grandfather wrote at great length in relation to the decomposed condition of the body.* He kept a copy of this letter, without any address, which I will give, omitting all about the body:

Lichfield, Feb. 7, 1767.

Dear Sir,

I am sorry you should think it necessary to make any excuse for a Letter I this morning received from you. The Cause of Humanity needs no Apology to me.

*   *   *   *

The Women that have committed this most unnatural crime, are real objects of our greatest Pity; their education has produced in them so much Modesty, or sense of Shame, that this artificial Passion overturns the very instincts of Nature!—what Struggles must there be in their minds, what agonies!—at a Time when, after the Pains of Parturition, Nature has designed them the sweet Consolation of giving Suck to a little helpless Babe, that depends on them for its hourly existence!

Hence the cause of this most horrid crime is an excess of what is really a Virtue, of the Sense of Shame, or Modesty. Such is the Condition of human Nature!

I have carefully avoided the use of scientific terms in this Letter that you may make any use of it you may think proper; and shall only add that I am veryly convinced of the Truth of every part of it.

and am, Dear Sir,

Your affectionate friend and servant,
Erasmus Darwin.

I give below two of his letters to Josiah Wedgwood.

ERASMUS DARWIN TO JOSIAH WEDGWOOD

Lichfield, Sept. 30, 1772.

Dear Wedgewood,

I did not return soon enough out of Derbyshire to answer your letter by yesterday's Post. Your second letter gave me great consolation about Mrs. Wedgewood, but gave me most sincere grief about Mr. Brindley, whom I have always esteemed to be a great genius, and whose loss is truly a public one. I don't believe he has left his equal. I think the various Navigations should erect him a monument in Westminster Abbey, and hope you will at the proper time give them this hint.

Mr. Stanier sent me no account of him, except of his death, though I so much desired it, since if I had understood that he got worse, nothing should have hindered me from seeing him again. If Mr. Henshaw took any Journal of his illness or other circumstances after I saw him, I wish you would ask him for it and enclose it to me. And any circumstances that you recollect of his life should be wrote down, and I will some time digest them into an Eulogium. These men should not die, this Nature denies, but their Memories are above her Malice. Enough!

*   *   *   *

ERASMUS DARWIN TO JOSIAH WEDGWOOD

Lichfield, Nov. 29, 1780.

Dear Sir,

Your letter communicating to me the death of your friend, and I beg I may call him mine, Mr. Bently, gives me very great concern; and a train of very melancholy ideas succeeds in my mind, unconnected indeed with your loss, but which still at times casts a shadow over me, which nothing but exertion in business or in acquiring knowledge can remove.

This exertion I must recommend to you, as it for a time dispossesses the disagreeable ideas of our loss; and gradually their impression or effect upon us becomes thus weakened, till the traces are scarcely perceptible, and a scar only is left, which reminds us of the past pain of the united wound.

Mr Bently was possessed of such variety of knowledge, that his loss is a public calamity, as well as to his friends, though they must feel it the most sensibly! Pray pass a day or two with me at Lichfield, if you can spare the time, at your return. I want much to see you; and was truly sorry I was from home as you went up; but I do beg you will always lodge at my house on your road, as I do at yours, whether you meet with me at home or not.

I have searched in vain in Melmoth's translation of Cicero's letters for the famous consolatory letter of Sulpicius to Cicero on the loss of his daughter (as the work has no index), but have found it, the first letter in a small publication called 'Letters on the most common as well as important occasions in Life', Newberry, St. Paul's, 1758. This letter is a masterly piece of oratory indeed, adapted to the man, the time, and the occasion. I think it contains everything which could be said upon the subject, and if you have not seen it I beg you to send for the book.

For my own part, too sensible of the misfortunes of others for my own happiness, and too pertinaceous of the remembrance of my own {i.e. the death of his son Charles in 1778}, I am rather in a situation to demand than to administer consolation. Adieu. God bless you, and believe me, dear Sir, your affectionate friend,

E. Darwin.

Ten years later he seems to have doubted much about the consolation to be derived from the letter of Sulpicius, for he writes (1790) to Edgeworth:[30]

> I much condole with you on your late loss. I know how to feel for your misfortune. The little Tale you sent is a prodigy, written by so young a person, with such elegance of imagination.
>
> Nil *admirari* may be a means to escape misery, but not to procure happiness. There is not much to be had in this world – we expect too much!
>
> I have had my loss also. The letter of Sulpicius to Cicero is fine eloquence, but comes not to the heart: it tugs, but does not draw the arrow. Pains and diseases of the mind are only cured by Forgetfulness [C.D. has 'Time']. Reason but skins the wound, which is perpetually liable to fester again.

The [excerpts from the] following letter may serve as a specimen of one of his speculative productions. Its date should be borne in mind in judging of its merits.

ERASMUS DARWIN TO JOSIAH WEDGWOOD

March, 1784.

Dear Sir,

I admire the way in which you support your new theory of freezing steam. You say, "Will not vapour freeze with a less degree of cold than water in the mass? instances hoar-frost, &c." Now this same et cætera, my dear friend, seems to me to be a gentleman of such consequence to your theory, that I wish he would unfold himself a little more.

I sent an account of your experiment to Mr. Robert, and desired him to show it to Dr. Black, so that I shall hope some time to hear his opinion on the very curious fact you mention, of a part of ice (during a thaw) freezing whilst you applied a heated body to another part of it.

Now in spite of your et cætera, I know no fact to ascertain that vapour will freeze with less cold than water.

*   *   *   *

I can in no way understand why, during the time you apply a heated body to one part of a piece of ice, when the air of your room was at 50° and the ice had for a day or two been in a thawing state, that a congelation should be formed on another part of the same ice, but from the following circumstances.

There is great analogy between the laws of the propagation of heat, and those of electricity, such as the same bodies communicate them easily, as metals, and the same bodies with more difficulty, as glass, wax, air: they are both excitable by friction, both give light, fuse metals, et cætera. Therefore I suppose that atmospheres of heat of different densities, like atmospheres of electricity, will repel each other at certain distances, like globules of quicksilver pressed against each other, and that hence by applying a heated body near one end of a cold body, the more distant end may immediately become warmer than the end nearest to the heated body.

*   *   *   *

March 11, 1784. Since I wrote the above I have reconsidered the matter, and am of opinion that steam, as it contains more of the element of heat than water, must require more absolute cold to turn it into ice, though the same sensible cold, as is necessary to freeze water, and that the phenomenon you have observed, depends on a circumstance which has not been attended to. When water is cooled down to freezing point, its particles come so near together, as to be within the sphere of their reciprocal attractions;— what then happens?—they accede with violence to each other and become a solid, at the same time pressing out from between them some air, which is seen to form bubbles in ice and renders the whole mass lighter than water (on which it will swim) by this air having regained its elasticity; and pressing out any saline matters, as sea-salt, or blue vitriol, which have become dissolved in it; and lastly by thus forcibly acceding together, the particles of water press out also some more heat, as is seen by the rising of the thermometer immersed in such freezing water.

This last circumstance demands your nice attention, as it explains the curious fact you have observed. When the heat is so far taken away from water, that the particles attract each other, they run together with violence, and press out some remaining heat, which existed in their interstices. Then the contrary must also take place when you add heat to ice, so as to remove the particles into their reciprocal spheres of repulsion: they recede from each other violently, and thence attract more heat into their interstices; and if your piece of hot silver is become cold, and has no more heat to give, or if this thawing water in this its expansile state is in contact with other water which is saturated with heat, it will rob it of a part, or produce freezing if that water was but a little above 32°.

I don't know if I have expressed myself intelligibly. I shall relate an experiment I made 25 years ago, which confirms your fact. I filled a brewing-copper, which held about a hogshead and half, with snow; and immersed about half-an-ounce of water at the bottom of a glass tube in this snow, as near the center as I could guess, and then making a brisk and hasty wood-fire under it, and letting the water run off by a cock as fast as it melted, I found in a few minutes on taking out the tube that the water in it was frozen. This experiment coincides with yours, and I think can only be explained on the above principle.

*   *   *   *

In support of the above theory I can prove from some experiments, that air when it is mechanically expanded always attracts heat from the bodies in its vicinity, and therefore water when expanded should do the same. But this would lengthen out my letter another sheet; I shall therefore defer it till I have the pleasure of a personal conference with you.

Thus ice in freezing gives out heat suddenly, and in thawing gives out cold suddenly; but this last fact had not been observed (except in chemical mixtures) because when heat has been applied to thaw ice, it has been applied in too great quantities.

When shall we meet? Our little boy has got the ague, and will not take bark, and Mrs. Darwin is therefore unwilling to leave him, and begs to defer her journey to Etruria till later in the season. Pray come this way to London or from London. Our best compts. to all yours,

Adieu,
E. Darwin.

P.S.—Water cooled beneath 32°, becomes instantly ice on any small agitation, or pouring out of one vessel into another, because thus the accession of the particles to each other, and the pressing out of the air, or saline matters, and of heat is facilitated.

Erasmus Darwin to his Son Robert

April 19, 1789.

Dear Robert,

I am sorry to hear you say you have many enemies, and one enemy often does [not?] much harm. The best way, when any little slander is told one, is never to make any piquant or angry answer; as the person who tells you what another says against you, always tells them in return what you say of them. I used to make it a rule always to receive all such information very coolly, and never to say anything biting against them which could go back again; and by these means many who were once adverse to me, in time became friendly. Dr. Small always went and drank tea with those who he heard had spoken against him; and it is best to show a little attention at public assemblies to those who dislike one; and it generally conciliates them.

*   *   *   *

My father seems to have consulted his father about some young man, whom he wished to see well started as an apothecary, and received the following answer:

Derby, Dec. 17, 1790.

Dear Robert,

I cannot give any letters of recommendation to Lichfield, as I am and have been from their infancy acquainted with all the apothecaries there; and as such letters must be directed to some of their patients, they would both feel and resent it. When Mr. Mellor went to settle there from Derby I took no part about him. As to the prospect of success there, if the young man who is now at Edinburgh should take a degree, (which I suppose is probable) he had better not settle in Lichfield.

I should advise your friend to use at first all means to get acquainted with the people of all ranks. At first a parcel of blue and red glasses at the windows might gain part of the retail business on market days, and thus get acquaintance with that class of people. I remember Mr. Green, of Lichfield, who is now growing very old, once told me his retail business, by means of his show-shop and many coloured window, produced him £100 a year.

Secondly, I remember a very foolish, garrulous apothecary at Cannock, who had great business without any knowledge or even art, except that he persuaded people he kept good drugs; and this he accomplished by only one stratagem, and that was by boring every person who was so unfortunate as to step into his shop with the goodness of his drugs. "Here's a fine piece of assafaetida, smell of this valerian, taste this album graecum. Dr. Fungus says he never saw such a fine piece in his life".

Thirdly, dining every market day at a farmers' ordinary would bring him some acquaintance, and I don't think a little impediment in his speech would at all injure him, but rather the contrary by attracting notice.

Fourthly, card assemblies,—I think at Lichfield surgeons are not admitted as they are here;—but they are to dancing assemblies; these therefore he should attend.

Thus have I emptied my quiver of the *arts* of the Pharmacopol. Dr. K—d, I think, supported his business by perpetual boasting, like a Charlatan; this does for a blackguard character, but ill suits a more polished or modest man.

If the young man has any friends at Shrewsbury who could give him letters of introduction to the proctors, this would forward his getting acquaintance.

For all the above purposes some money must at first be necessary, as he should appear well; which money cannot be better laid out, as it will pay the greatest of all interest by settling him well for life. Journeymen Apothecaries have not greater wages than many servants; and in this state they not only lose time, but are in a manner lowered in the estimation of the world, and less likely to succeed afterwards.

I will certainly send to him when first I go to Lichfield. I do not think his impediment of speech will injure him; I did not find it so in respect to myself. If he is not in such narrow circumstances but that he can appear well, and has the knowledge and sense you believe him to have, I dare say he will succeed anywhere.

A letter of introduction from you to Miss Seward, mentioning his education, may be of service to him, and another to Mr. Howard.

> Adieu from, dear Robert,
> Yours most affectionately,
> E. Darwin.

His activity continued to his latest days; and the following letter to my father, written when he was sixty-one years old, shows his continued zeal in his profession.

ERASMUS DARWIN TO HIS SON ROBERT

Derby, April 13, 1792.

Dear Robert,

I think you and I should sometimes exchange a long medical letter, especially when any uncommon diseases occur; both as it improves one in writing clear intelligible English, and preserves instructive cases. Sir Joshua Reynolds in one of his lectures on pictorial taste, advised painters, even to extreme old age, to study the works of all other artists, both ancient and modern; which he says will improve their invention, as they will catch collateral ideas (as it were) from the pictures of others, which is a different thing from imitation; and adds, that if they do not copy others, they will be liable to copy *themselves*, and introduce into their work the same faces, and the same attitudes again and again. Now in medicine I am sure unless one reads the work of others, one is liable perpetually to copy one's *own* prescriptions, and methods of treatment; till one's whole practice is but an imitation of one's self; and half a score medicines make up one's whole materia medica; and the apothecaries say the doctor has but 4 or 6 prescriptions to cure all diseases.

Reasoning thus I am determined to read all the new medical journals which come out, and other medical publications, which are not too voluminous; by which one knows what others are doing in the medical world, and can astonish apothecaries and surgeons with the new and wonderful discoveries of the times. All this harangue lately occurred to me on reading the trials made by Dr. Crawford.

*   *   *   *

Amongst the old letters preserved, there is one without any date from James Hutton, the founder of the modern science of geology, and I extract its commencement, as proceeding from so illustrious a scientific man. My grandfather seems to have complained to him of having been cheated by some publisher; and Hutton answers:

> If you have no more money than you use, then be as sparing of it as you please, but if you have money to spend, then pray learn to let

yourself be cheated, that is, learn to lay out money for which you have no other use. If this be not philosophy, at least it is good sense; for why the devil should a man have money to be a plague to him, when it is so easy to throw it away; and if thro' a spirit of general benevolence you are afraid of mankind suffering from this root of all evil, for God's sake send it to the bottom of the sea, it there can only poison fish and it will there make in time a noble fossil specimen.

Though my grandfather lived through a most exciting period of history, it is singular how rarely there is more than an allusion in his letters to politics. He was what would now be called a liberal, or perhaps rather a radical. He seems to have wished for the success of the North American colonists in their war for independence; for he writes to Wedgwood (Oct. 17, 1782), "I hope Dr. Franklin will live to see peace, to see America recline under her own vine and fig-tree, turning her swords into plough-shares, &c."

Like so many other persons, he hailed the beginning of the French Revolution with joy and triumph. Miss Seward in a letter to Dr. Whalley, dated May 18, 1792, says: "I should indeed now begin to fear for France; but Darwin yet asserts that, in spite of all their disasters, the cause of freedom will triumph, and France become, ere long, an example, prosperous as great, to the surrounding nations".

She remarks in another letter that Darwin "was a far-sighted politician, and foresaw and foretold the individual and ultimate mischief of every pernicious measure of the late Cabinet".[31]

———————————— ⬲ ————————————

My father spoke of his father as having great powers of conversation; and he told me that Lady Charleville, who had been accustomed to the most brilliant society in London, said that Dr. Darwin was one of the most agreeable men whom she had ever met. He himself used to say "there were two sorts of agreeable persons in conversation parties—agreeable talkers and agreeable listeners". Miss Seward speaks of him as being extremely sarcastic, but of this I can find no evidence in his letters or elsewhere.

It is a pity that Dr. Johnson in his visits to Lichfield rarely met Dr. Darwin; but they seem to have disliked each other cordially, and to have felt that if they met they would have quarrelled like two dogs. There can, I suppose, be little doubt that Johnson would have come off victorious. In a volume of MSS. by Dr. Darwin, in the possession of one of his granddaughters, there is the following stanza, on Mr. Seward's edition of 'Beaumont and Fletcher's Plays', and on Dr. Johnson's edition of 'Shakespear's Plays', 1765:

> From Lichfield famed two giant critics come,
> Tremble ye Poets! hear them! "Fe, Fo, Fum!"
> By Seward's arm the mangled Beaumont bled,
> And Johnson grinds poor Shakespear's bones for bread.

He possessed, according to my father, great facility in explaining any difficult subject; and he himself attributed this power to his habit of always talking about whatever he was studying, "turning and moulding the subject according to the capacity of his hearers". He compared himself to Gil Blas's uncle, who learned the grammar by teaching it to his nephew.

When he wished to make himself disagreeable for any good cause, he was well able to do so. Lady * * * married a widower, and became so jealous of his former wife that she cut and spoiled her picture, which hung up in one of the rooms. The husband was greatly alarmed, and sent for Dr. Darwin, fearing that his young wife was becoming insane. When he arrived he told her in the plainest manner many unpleasant truths, amongst others that the former wife was infinitely her superior in every respect, including beauty. The poor lady was astonished at being thus treated, and could never afterwards endure my grandfather's name. He told the husband if she ever again behaved oddly, to hint that he would be sent for. The plan succeeded perfectly, and she ever afterwards restrained herself.

As my father was separated so early in life from his father, he remembered few of his remarks, but he used to quote one saying as very true, according to his own experience, viz., "that the world was not governed by the clever men, but by the active and energetic". He used also to quote another saying, that "common sense would be improving, when men left off wearing as much flour on their heads as would make a pudding; when women left off wearing rings in their ears, like savages wear nose rings; and when fire-grates were no longer made of polished steel". *If he had lived to the present day he would have agreed that the world had grown somewhat wiser.*

Here may be given his free and happy translation of an epigram of Martial, written probably whilst young:[32]

> Balnea, Vina, Venus corrumpunt corpora nostra,
> At faciunt vitam Balnea, Vina, Venus.
> Wine, women, warmth, against our lives combine,
> But what is life without warmth, women, wine!

He stammered greatly, and it is surprising that this defect did not spoil his powers of conversation. A young man once asked him in, as he thought, an offensive manner, whether he did not find stammering very inconvenient. He answered, "No, Sir, it gives me time for reflection, and saves me from asking impertinent questions".

Manners seem to have been very coarse in those old days. A notorious man, Dr. Caleb Hardinge, who seems to have lived in good society, said to my grandfather, "my dear doctor, you have a damned ugly trick of stuttering; I am sure I could cure you". To which my grandfather replied, alluding, evidently to the swearing, "Physician heal thyself".[33]

We may here pause, and endeavour to weigh his intellectual powers. Judging from his published works, letters, and all that I have been able to gather about him, the vividness of his imagination seems to have been one of his pre-eminent characteristics. This led to his great originality of thought, his prophetic spirit both in science and in the mechanical arts, and to his overpowering tendency to theorise and generalise. Nevertheless, his remarks (previously given) on the value of experiments and the use of hypotheses show that he had the true spirit of a philosopher. That he possessed uncommon powers of observation must be admitted. The diversity of the subjects to which he attended is surprising. But of all his characteristics, the incessant activity or energy of his mind was, perhaps, the most eminent.

Mr Keir, himself a distinguished man, who had seen much of the world, and who "had been well acquainted with Dr. Darwin for nearly half a century", after his death wrote (May 12, 1802) to my father, and I will give a few extracts from this long letter confirming my own independent judgment. After remarking how "our actions are regulated by custom, and the eccentricities of our characters are rounded by the collision of society", Keir says: "Your father did indeed retain more of his original character than almost any man I have known, excepting, perhaps, Mr. Day {author of 'Sandford and Merton', &c.}. Indeed, the originality of character in both these men was too strong to give way to the example of others". He afterwards proceeds:

> Your father paid little regard to authority, and he quickly perceived the analogies on which a new theory could be founded. This penetration or sagacity by which he was able to discover very remote causes and distant effects, was the characteristic of his understanding. Perhaps it may be thought in some instances to have led him to refine too much, as it is difficult in using a very sharp-pointed instrument to avoid sometimes going rather too deep. By this penetrating faculty he was enabled not only to trace the least conspicuous indications of scientific analogy, but also the most delicate and fugitive beauties of poetic diction. If to this quality you add an uncommon activity of mind and facility of exertion, which required the constant exercise of some curious investigation, you will have, I believe, his principal features.

His activity continued to his latest days. My father seems to have urged him, about the year 1793, to leave off professional work. He answered:

It is a dangerous experiment and generally ends either in drunkenness or hypochondriacism. Thus I reason, one must do something (so country squires fox-hunt), otherwise one grows weary of life, and becomes a prey to ennui. Therefore one may as well do something advantageous to oneself and friends, or to mankind, as employ oneself in cards or other things equally insignificant.

During his frequent and long journeys, he read and wrote much in his carriage, which was fitted up for the purpose. Nor was travelling an easy affair in those days, for owing to the state of the roads a carriage could hardly reach some of the houses which he had to visit; and I hear from one of his granddaughters that "an old horse named the 'Doctor', with a saddle on, used to follow behind the carriage, without being in any way fastened to it; and when the road was too bad, he got out and rode upon Doctor. This horse lived to a great age, and was buried at the Priory."

When at home he was an early riser; and he had his papers so arranged (as I have heard from my father) that if he awoke in the night he was able to get up and continue his work for a time, until he felt sleepy. Considering his indomitable activity, it is a singular fact that he suffered much from a sense of fatigue. This I likewise heard from my father, on my remarking to him how greatly fatigued he seemed to be after his day's work, and he answered, "I inherit it from my father".

Dr. Darwin has been frequently called an atheist, *apparently as a convenient term of abuse*; whereas in every one of his works distinct expressions may be found showing that he fully believed in God as the creator of the universe. For instance, in the 'Temple of Nature', published posthumously,[34] he writes: "Perhaps all the productions of nature are in their progress to greater perfection! an idea countenanced by modern discoveries and deductions concerning the progressive formation of the solid parts of the terraqueous globe, and consonant to the dignity of the Creator of all things". {See also the striking footnote (p.142) on the immutable properties of matter, 'received from the hand of the Creator'.} He concludes the section on Generation in the 'Zoonomia' with the words of the Psalmist: "The heavens declare the Glory of God, and the firmament sheweth his handiwork".

He also wrote as a college exercise [C.D. has 'also published'] an ode on the folly of atheism, with the motto "I am fearfully and wonderfully made". I have seen two copies, which differ slightly, and will give the first and last stanzas of one of them:

I.

Dull atheist, could a giddy dance
Of atoms lawless hurl'd
Construct so wonderful, so wise,
So harmonised a world?

14.

What Potent-power, all-great, all-good
Do these around me own?
Teach me, Creation, teach me how
T' adore the vast Unknown!

A hymn attributed to Dr. Darwin likewise appeared in a collection edited by Dr. Kippis, in 1795, and which has often been republished. The Rev. J. Martineau doubted whether the hymn was really written by Dr. Darwin, but a copy was found with his initials in the sixth volume of his shorthand notes; and as he did not long continue using shorthand, the hymn must have been written whilst he was rather young. I will give the first and last verses; it is headed as follows:—"Prosperity and adversity, life and death, poverty and riches come of the Lord.—Ecc. ii., 14".

I.

The Lord! how tender is His love,
His justice how august!
Hence, all her fears my soul derives,
There, anchors all her trust.

7.

Oh grant that still with grateful heart
My years resign'd may run;
'Tis Thine to give or to resume,
And may Thy will be done.

Although Dr. Darwin was certainly a theist in the ordinary acceptation of the term, he disbelieved in any revelation. {I have heard from a connection of the family that an old and faithful maid-servant, who was present when my grandfather died, afterwards told his step-daughter, that on hearing him faintly saying something, she bent down her head to listen, but the only word she caught was "Jesus". This was an inexpressible comfort to his step-daughter (from whose daughter I have received this account), and she spoke of it only a few weeks before her own death. But it is incredible that my grandfather should have wholly changed his judgment on so important a subject, and that my father should never have heard of it, though he hastened to Derby on the news of his father's death. The good old maid-servant must have fancied that she heard what she wished to hear; and the statement may be added to the many apocryphal death-scenes on record. That comfort should be derived by relations from a change in the settled convictions of a life-time, just when sense and reason are failing, is a strange fact, but seems to be a part of human nature.}

Nor did he feel much respect for unitarianism, for he used to say that "unitarianism was a feather-bed to catch a falling Christian".

———————— ⌒ ————————

As it is the duty of a biographer to draw as true a picture as he can of whomsoever he may describe, I will mention such blemishes in the character *of my grandfather as I have been able to discover, before giving the favourable side of the picture.*

From my father's conversation, I infer that his father had acted towards him in his youth rather harshly and imperiously, and not always justly; and though my grandfather in after years felt the greatest interest in his son's success, and frequently wrote to him with affection, in my opinion the early impression on my father's mind was never quite obliterated. *Although my father often expressed the highest admiration of his father's genius, I doubt whether he felt any strong affection for him; but another member of the family differs from me in this respect.*

I have heard indirectly (through one of his stepsons) that he was not always kind to his son Erasmus, being often vexed at his retiring nature, and at his not more fully displaying his great talents. His two younger sons [Erasmus and Robert] differed widely from him in their tastes, and this may for a time have rendered him less kind to them. His children by his second marriage seem to have entertained the warmest affection for him.

In the interval between his first and second marriages, Dr. Darwin became the father of two illegitimate daughters. To his credit be it said that he gave them an excellent education, and from all that I have heard they grew up to be admirable ladies, living on intimate terms with his widow and the children by the second marriage.

In our present state of society it may seem a strange fact that my grandfather's practice as a physician should not have suffered by his openly bringing up illegitimate daughters. *But we should remember that about this same time Lord Thurlow, the Chancellor and Keeper of the King's conscience, "lived openly with a mistress, and had a family by her, whom he recognised and without any disguise brought out in society as if they had been his legitimate children".*[35] {At a rather later period, according to Lord Campbell, "the understanding was that a man elevated to the Bench, if he had a mistress, must either marry her, or put her away".}

We now come to a less grave charge. Miss Seward accuses my grandfather of having appropriated several verses from her, altering some and adding others, and published them in the 'Botanic Garden' without any acknowledgment. The case is a very odd one; for firstly, she herself admits[36] that it was entirely through his instrumentality that these verses were published with her name attached to them, before the appearance of the 'Botanic Garden', in the 'Monthly Magazine', and afterwards in the 'Annual Register'. Secondly, there seems to have been little temptation for the theft, for the whole history of his life shows that writing verses on any subject was not the least labour to him, but only a pleasure. And thirdly, that Miss Seward remained on the same friendly, almost playful, terms with him afterwards as before. The whole case is unintelligible, and in some respects looks more like highway robbery or the exaction of blackmail than simple plagiarism.

Mr Edgeworth, in a letter (Feb. 3, 1812) to Sir Walter Scott,[37] says that he had expressed surprise to Dr. Darwin at seeing Miss Seward's lines at the beginning of his poem, and that Dr. Darwin replied: "It was a compliment which he thought himself bound to pay to the lady, though the verses were not of the same tenor as his own". But this seems a lame excuse, and it is an odd sort of compliment to take the verses without any acknowledgment. Perhaps he thought it fair play, for Edgeworth goes on to say that

> Miss Seward's 'Ode to Captain Cook' stands deservedly high in public opinion. Now to my certain knowledge most of the passages which have been selected in the various reviews of the work were written by Dr. Darwin. . . . I knew him well, and it was as far from his temper and habits, as it was unnecessary to his acquirements, to beg, borrow, or steal from any person on earth.

These passages anyhow show how true and ardent a friend Edgeworth was to Dr. Darwin long after his death.

———————————————— ⌒ ————————————————

We will now turn to the favourable side of his character; and it will save some repetition if we consider in the order of publication the many false statements and calumnies which have appeared about him, discussing at the same time any related subjects. This plan has, however, the disadvantage of rendering it impossible to arrange the several moral qualities in any natural order.

Shortly after the death of Dr. Darwin, a long article appeared,[38] reviewing, for the most part favourably his character and works, with some just criticisms of the latter. But towards the close it is said:

> There was one great end, to the attainment of which all his talents and views were earnestly and uniformly directed. He did not hesitate openly and repeatedly to declare in public company, that the acquisition of wealth was the leading object of all his literary undertakings! He once said to a friend, "I have gained £900 by my 'Botanic Garden' and £900 by the first volume of 'Zoonomia', and if I can every other year produce a work which will yield this sum I shall do very well". He added, "Money and not fame is the object which I have in view in all my publications".
>
> But Dr. Darwin was by no means insensible to the value of reputation. During the last years of his life, the love of fame was a passion which had great power over his mind; and the incense of praise was so pleasant to him, that flattery was found to be the most successful means of gaining his notice and favour.

All that I have been able to learn goes to show that this was a mistaken view of his character. *I can positively declare that I have met with nothing in his many letters, or heard of any anecdote, showing an inordinate love of wealth, or anxiety about fame; to have been altogether indifferent about money would have been culpable, seeing that he had a large family to support.*

In a letter to my father, dated Feb. 7, 1792, he writes:

> I always console myself by saying, "a man must do something, otherwise he becomes moped and low-spirited, and therefore he may as well do something for the advantage of himself and family, as labour in fox-hunting or card-playing". So life rubs on! As to fees, if your business pays you well on the whole, I would not be uneasy about making absolutely

the most of it. To live comfortably all one's life, is better than to make a very large fortune towards the end of it.

In another letter not dated, but written in 1793, he remarks: "There are two kinds of covetousness, one the fear of poverty, the other the desire of gain. The former, I believe, at some times affects all people who live by a profession". Again, his son Erasmus, in writing on Nov. 12, 1792 to my father, after remarking how rich he was becoming, adds: "I am not afraid of being rich, as our father used to say at Lichfield he was, for fear of growing covetous; to avoid which misfortune, as you know, he used to dig a certain number of duck puddles every spring, that he might fill them up again in the autumn". How it was possible to expend much money in digging duck puddles, whatever these may have been, it is not easy to see.

Now is it likely that a man who habitually spoke and wrote in the above manner to his own family, should in earnest have said, or rather boasted, that he cared only for money? There is indeed abundant evidence that he was charitable, generous, and extremely hospitable.

It is probable that the above statements of the reviewer, and of several other statements, originated in his habit—perhaps a foolish one—of often speaking about himself in a quizzing or bantering tone. Mr. Edgeworth, who had known him "intimately during thirty-six years", in answer to the reviewer, writes:[39]

> *I am sufficiently acquainted with the world to know that one foolish friend is more dangerous than twenty enemies; I therefore abstained from the precipitate expression of that indignation which every honest mind must feel when the character of a great and good man is disfigured by misrepresentation or ridicule.*

He then proceeds:

> I am most anxious to contradict that assertion of the anonymous biographer, which I consider the most unfounded and injurious—that Dr. Darwin wrote chiefly for money.... It is not improbable that, to avoid offensive adulation, he might have said

ironically that his object in writing was money, not fame. I have heard him say so twenty times, but I never for one moment supposed him to be in earnest. *Indeed it was absolutely impossible that I should. I once, when in England, had a sudden occasion for a thousand pounds. Knowing that the doctor had money in his banker's hands, I wrote to him to request that he would within a fortnight accommodate me with that sum for a few weeks. By return of post I received the following answer: 'I send you one bank-note for £1000. Send me a bond secundum artem'. The Doctor at that time knew nothing of my affairs, but he thought me worthy to be his friend.*

Mr. Edgeworth then proceeds:

It is asserted by the reviewer 'that he stooped to accept of gross flattery'. Perhaps in the inmost recesses of his heart, vanity might reign without control, but no man exacted less tribute of applause in conversation. When the admirable *travestie* of his poetic style was published in the Anti-jacobin newspaper, I spoke of it in his presence in terms of strong approbation, and he appeared to think as I did, of the wit, ingenuity, and poetic merit of the parody.

To have asked the author of the 'Loves of the Plants' to admire the 'Loves of the Triangles' was putting his temper through a severe ordeal.

Mr. Keir, who had known Dr. Darwin well for nearly half a century, remarks in a letter (May 12, 1802) before the appearance of the above review: "The works of your father are a more faithful monument and more true mirror of his mind than can be said of those of most authors. For he was not one of those who wrote *invitâ Minervâ*, or from any other incitement than the ardent love of the subject".

Throughout his letters I have been struck with his indifference to fame, and the complete absence of all signs of any over-estimation of his own abilities or of the success of his works. I infer, but only from his having mentioned the fact to my father, that he was pleased by receiving what he describes as "a portrait of my head, well done, I believe—proofs, 10s. 6d.—the first impression of which the engraver, Mr. Smith, believes will soon be sold, and he will then sell a second at 5s." My grandfather adds: "but the great honour of all is to have one's head upon a sign post, unless, indeed, upon Temple Bar!" This engraving was copied from the picture by Wright, of which a photograph is given in the present volume.

Many pictures were made of him, but with one or two exceptions they are characterised by a rather morose and discontented expression. Mr. Edgeworth, in writing to my grandfather about one of these pictures, says: "There is a cloud over your brow and a compression of the lips that hide your benevolence and good humour. And great author as you are, my dear doctor, I think you excel the generality of mankind as much in generosity as in abilities."[40]

I have said that, as far as I can judge, he was remarkably free from vanity, conceit, or display; nor does he appear to have been ambitious for a higher position in society. Lady Charlotte Finch, governess to Queen Charlotte's daughters, had two granddaughters, the Miss Fieldings, one of whom was taken to Dr. Darwin, at Derby, on account of her health, and was invited to stay some time at his house. George the Third heard of my grandfather's fame (as I have been informed by one of his granddaughters) through Lady Charlotte, and said: "Why does not Dr. Darwin come to London? He shall be my physician if he comes"; and he repeated this in his usual manner. But Dr. Darwin and his wife agreed that they disliked the thoughts of a London life so much, that the hint was not acted on. Others have expressed surprise that he never migrated to London.

There is one other statement in the 'Monthly Magazine', which must be noticed. The reviewer affirms that just before he died he "fell into a violent fit of passion with his servant", and that this probably hastened his death. His two medical attendants, Dr. Fox and Mr. Hadley, immediately published a notice that this statement was wholly false. If, indeed, the old maid-servant, whose apocryphal account of the last word which he uttered has *already been given, can be trusted—and I know of no reason why she should not be trusted—his last action, an hour before he died, was one of considerate kindness.* Whilst writing in the early morning a long and affectionate letter to Mr. Edgeworth, he was seized with a violent shivering fit, and went into the kitchen to warm himself before the fire. He there saw this maid churning, and asked her why she did this on a Sunday morning. She answered that she had always done so, as he liked to have fresh butter every morning. He said: "Yes, I do, but never again churn on a Sunday!"

That he was irascible there can be no doubt. My father says "he was sometimes violent in his anger, but . . . his sympathy and benevolence soon made him try to soothe or soften matters". Mr Edgeworth also says,[41] in reference to the above accusation:

> Five or six times in my life I have seen him angry, and have heard him express that anger with much real, and more apparent vehemence—more than men of less sensibility would feel or show. But then the motive never was personal. When Dr. Darwin beheld any example of inhumanity or injustice, he never could restrain his indignation; he had not learnt, from the school of Lord Chesterfield, to smother every generous feeling.

———————————— ⇝ ————————————

In 1804 Miss Seward published her 'Life of Dr. Darwin'. This was an unfortunate event for his good fame, for she knew nothing about science or medicine, and her style is so pretentious that it is extremely disagreeable, almost nauseous to many persons; though others like the book much. It abounds with inaccuracies, as both my father and other members of the family asserted at the time of its publication. For instance, she states that when dying he sent for Mrs. Darwin, and first asked her and then his daughter Emma to bleed him, and gives their answers in inverted commas. But the whole account is a simple fiction, for he expressly told his servant not to call Mrs. Darwin, but was disobeyed by the servant, who saw how ill he was; and his daughter was not even present. She does not even give his age correctly at the time of his death. It is also obvious that the many long speeches inserted in her book are the work of her own imagination, either with some or with no foundation.

She describes (p. 406) my grandfather's conduct when he heard of his son's suicide as brutal to an unparalleled degree. She asserts that when he was told that the body "was found, he exclaimed in a low voice, 'Poor insane coward', and, it is said, never afterwards mentioned the subject". Miss Seward then proceeds (p. 408):

> This self-command enabled him to take immediate possession of the premises bequeathed to him (by his son Erasmus); to lay plans for their improvement; to take pleasure in describing those plans to his acquaintance, and to determine to make it his future residence; and all this without seeming to recollect to how sad an event he owed their possession!

The whole of this account is absolutely false, and when my father demanded her authority, she owned that it had been given merely on a report at a distant place, without any inquiry having been made from a single person who could have authentically known what happened. On the day after the death of his son (December 30th, 1799), my grandfather wrote to my father, and it appears that he at first either supposed that the drowning was accidental, or he wished to spare my father's feelings; for he says: "I write in great anguish of mind to acquaint you with a dreadful event—your poor brother Erasmus fell into the water last night at the bottom of his garden, and was drowned". When the news was brought to him by Mr. Parsons (one of Erasmus's clerks) that the body had been at last found, two of his daughters, Emma and Violetta were with him, and the former gave the following account to my mother:

> He immediately got up, but staggered so much that Violetta and I begged of him to sit down, which he did, and leaned his head

upon his hand . . . he was exceedingly agitated, and did not speak for many minutes. His first words were, 'I beg you will not, any of you, ask to see your poor brother's corpse'; and upon our assuring him that we had not the least wish to do so, he soon after said that this was the greatest shock he had felt since the death of his poor Charles.

Emma then asserts that Miss Seward's other statements are utterly false, namely, that he never afterwards mentioned his son's death, and that he took immediate possession of the property bequeathed to him. After alluding to other inaccuracies in Miss Seward's book, Emma concludes in a truly feminine and filial spirit:

> There is nothing else of such infinite consequence as her daring publicly to accuse my dear papa of want of affection and feeling towards his son. How can this be contradicted? I want to scratch a pen over all the lies, and send the book back to Miss Seward; but mamma won't allow this. She thinks you and my brother will think of a better plan; for myself, I should feel no objection to swear the truth of what I have said before both Houses of Parliament.

I will only add that, according to my father (in a letter to Miss Seward, Feb. 10, 1804): "My father, even to the last months of his life, said that it required constant, earnest, and forced exertion to study to obtain relief, and he did actually pursue certain studies for this sole object". In one of my grandfather's letters, dated Feb. 8, 1800, he writes: "I am obliged as executor daily to study his (Erasmus's) accounts, which is both a laborious and painful business to me". A fortnight afterwards he tells my father about a monument to be erected to Erasmus, and adds: "Mrs Darwin and I intend to lie in Breadsal church by his side".

Rarely has a greater calumny been published about anyone than the above account given by Miss Seward of the behaviour of my grandfather when he heard of his son's death.

{Miss Seward published, on my father's demand, the following retractation in several journals, but such retractations are soon forgotten, and the stigma remains:

The authoress of the 'Memoirs of Dr. Darwin', since they were published, has discovered, on the attestation of his family and other persons present at the juncture, that the statement given of his exclamation, p. 406, on the death of Mr. Erasmus Darwin, is entirely without foundation; and that the doctor, on that melancholy event, gave amongst his own family, proofs of strong sensibility at the time, and of succeeding regard to the memory of his son, which he seemed to have a pride in concealing from the world. In justice to his memory, she is desirous to correct the misinformation she had received.

('Monthly Magazine', 1804, p. 378; and other journals and newspapers.)}

That the act of suicide was committed during temporary insanity there can be little doubt. It is known that a change of disposition generally precedes insanity, and Erasmus, from being an excellent man of business, had become dilatory to an abnormal degree. It appears that he had neglected to do something of importance for my father; and my grandfather nearly two years before Erasmus's death, wrote in his excuse to my father (June 8, 1798) as follows:

I have not spoken to him on your affairs—his neglect of small businesses (as he thinks them, I suppose) is a constitutional disease. I learnt yesterday that he had like to have been arrested for a small candle bill of 3 or 4 pounds in London, which had been due 4 or 5 years, and they had in vain repeatedly written to him! and that a tradesman in this town has repeatedly complained to a friend of his that he owes Mr. D. £70, and cannot get him to settle his account. I write all this to shew you that his neglectful behaviour to you was not owing to any disrespect, or anger, or resentment, but from what?—from *defect of voluntary power.* Whence he procrastinates for ever!

The immediate cause of his death was the attempt to settle some accounts. His confidential clerk told one of Dr. Darwin's stepsons that Mr. Darwin had been working for two nights, and when urged in the evening of December 29th to take some rest and food, he answered with a most distressed expression, holding his head, "I cannot, for I promised if I'm alive that the accounts should be sent in to-morrow". Early in the night of the same day he could bear his misery no longer, and seems to have rushed out of the house and, leaving his hat on the bank, to have thrown himself into the water.

He was probably conscious of some mental change, for he purchased, six weeks before his death, some land near Lincoln, and the small estate of the Priory, near Derby; and he intended, at the early age of forty, to retire from business, so as to spend the rest of his days in quiet; or, as my grandfather, who could not have foreseen what all this foreboded, expressed it (in a letter to my father, Nov. 28, 1799), "to sleep away the remainder of his life".

Amongst the property of Erasmus my grandfather found a little cross made of platted grass (now in my possession), gathered from the tomb of Charles [Erasmus's elder brother], who had died twenty years before. A week before my grandfather died, he sent this to my father to be preserved.

The false reports about my grandfather's conduct on the death of his son, probably originated in his strong dislike to affectation, or to any display of emotion in a man. He therefore probably wished to conceal his own feelings, and perhaps did so too effectually. My father writes: "He never would allow any common acquaintance to converse with him upon any subject that he felt poignantly.... It was his maxim, that in order to feel cheerful you must appear to be so". There was, moreover, a vein of reserve in him, independently of any deep feeling; and Miss Seward, in answer to a remark by my father, says (May 10, 1802, i.e., before the publication of the 'Memoirs'):

> Too well was I acquainted with the disposition and habits of your
> lamented father, to feel surprise from your telling me how little

you had been able to gather from himself concerning the circumstances of his life, which preceded your birth, and those which passed beneath the unobservant eyes of sportive infancy.

The many friends and admirers of Dr Darwin were indignant at Miss Seward's book, and thought that it showed much malice towards him. No such impression was left on my mind when lately re-reading it, but only that of scandalous negligence, together, perhaps, with a wish to excite attention to her book, by inserting any wild and injurious report about him. The friends, however, of Dr. Darwin were right, for in a letter, dated May 12, 1802 (before she published the 'Memoirs'), and addressed to the Rev. Dr. Whalley,[42] she shows her true colours, and gives an odious character of "that large mass of genius and sarcasm", as she calls him:

> The deceased Dr. Darwin was a mixed character, illustrious by talent, and professionally liberal during his residence in Lichfield; always hospitable, sometimes friendly, but never amiable. While on his entrance into company, and on every commencing conversation, his countenance wore the open and exhilarating smile of benevolence, yet in the progress of that conversation, though his imagination glowed on abstracted themes, was there invariably found a cold satiric atmosphere around him, repulsing the confidence and sympathy of friendship. Age did not improve his heart; and on its inherent frost, poetic authorism, commencing with him after middle life, engrafted all its irritability, disingenuous arts, and grudging jealousy of others' reputation in that science. He had not the smallest confidence in human testimony, however respectable, if it was any way hostile to his theories.

Other charges are then added. We shall presently see how utterly false it is to say that he repulsed friendship, and was wanting in sympathy.

It is natural to inquire why Miss Seward felt so much malice towards a man with whom she had lived on intimate terms during many years, and for whom she often expressed, and probably felt the highest admiration. The explanation appears to be that

Dr. Darwin rejected her love. Even before his first marriage there appear to have been some love-passages between them, for Edgeworth remarks: "It seems that Mrs. Darwin had a little pique against Miss Seward, who had in fact been her rival for the doctor". According to my father and to several other members of the family, she again much wished to marry my grandfather before his second marriage; and I understood from my father that he possessed documentary evidence (subsequently destroyed) to this effect. This explains the following significant sentence in a letter written to her by my father on March 5th, 1804, in relation to her account of the suicide of Erasmus: "Were I to have published my father's papers in illustration of his conduct, some circumstances must unavoidably have appeared, which would have been as unpleasant for you to read as for me to publish". Disappointed affection, with some desire for revenge, render intelligible her whole course of conduct.

——————————— ⟫ ———————————

We now come to the calumnies published in 1858 in the 'Life' of Mrs. SchimmelPenninck,[43] who was the eldest sister of Tertius Galton, Dr. Darwin's son-in-law. These are hardly worth notice, as they were dictated in old age, she having seen Dr. Darwin, according to her own statement, only "with the eyes of a child". Nor was she a trustworthy person, at least when young, as I have been assured by her near relations, the several children of her brother, Tertius Galton. I have a copy of a letter written (Feb. 20, 1871) by one of Mrs. SchimmelPenninck's nieces to Dr. Dowson, who had used her book in his life of Dr. Darwin, and nothing can be more explicit than the statements about her untrustworthiness. For instance, "she had the habit of colouring her facts till they almost ceased to be true". She quarrelled with the members of her own family in 1810, "and her feelings towards them, which she shewed on various occasions, had evidently influenced her in her description of Dr. Darwin". A sister also, of Mrs SchimmelPenninck in speaking of the statements in her book, says: "They are facts distorted, and give a false impression".

She [Mrs SchimmelPenninck] *praises highly his powers of conversation, and in speaking of the wonderful set of men (chiefly members of the famous Lunar Club), who often met at her father's house, she says that* Dr. Darwin *"was the culminating point". But she accuses him of maintaining that "conscience and sentiment are mere figments of the imagination", that he did not care whether what he had written about the* Upas-tree *was true or not, and that he was a coarse glutton".*

Her condemnation was, however, chiefly directed against his avowed disbelief in revelation—in the existence of a God, the soul, and of a future life. Yet she admits, in reference to such subjects, that "his conversations, though extreme at that time, were yet in keeping with the universal spirit of the day". The explanation of much of what she says probably lies in her having mistaken, as a child, exaggerated statements about himself (formerly alluded to) made in joke as made in earnest. The granddaughters of Mr. and Mrs. Galton assure me that their grand-parents always spoke of Dr. Darwin with veneration and affection; and is it credible that widely-known and eminently respectable members of the Society of Friends would have tolerated a man who continually scoffed at conscience, truth, morality, and religion?

It is hardly worth while to notice any further Mrs. SchimmelPenninck's accusations, yet I will do so briefly. With respect to gluttony, I can only say that I never heard my father make any remark countenancing such an idea; nor have the Galtons heard anything of the kind from their mother; but she, unfortunately, never read the book, as her daughters knew how much it would have pained her. Dr. Darwin was a tall, bulky man, and lived largely on cream, fruit, and vegetables; it is therefore, probable that he ate largely, as every man must do who works hard, and lives on such a diet.

We have already seen how false it is to call him an atheist. It is manifestly untrue that he was indifferent about the truth of what he wrote on the Upas-tree, for he gives his authority in full detail. Is it likely that a man who disbelieved in conscience and morality would have written the following verses on Slavery?[44]

Throned in the vaulted heart, his dread resort,
Inexorable CONSCIENCE holds his court;
With still small voice the plots of Guilt alarms.
Bares his mask'd brow, his lifted hand disarms;
But wrapp'd in might with terrors all his own,
He speaks in thunder, when the deed is done.
Hear him, ye Senates! hear this truth sublime,
He, who allows oppression, shares the crime.

With reference to morality, he says:[45]

The famous sentence of Socrates, "Know thyself", ... however
wise it may be, seems to be rather of a selfish nature.... But the
sacred maxims of the author of Christianity, "Do as you would be
done by", and "Love your neighbour as yourself", include all our
duties of benevolence and morality; and, if sincerely obeyed by all
nations, would a thousandfold multiply the present happiness of
mankind.

In a Guide-book to Derbyshire published not many years ago, the
anonymous author plainly hints, but without giving any evidence, that
my grandfather had been guilty of murder. This man knew that he might
safely slander the dead. As the old friend of Dr. Darwin, Mr. Keir, wrote
(June 18, 1802) to my father after the publication of the first slanderous
attack: "It is not easy for good-natured people to conceive how far the
malignity and envy of mankind will go.... There never was, and, I believe,
never will be, any eminent man who was not the object of envy and abuse".

We have now finished with calumnies, and may turn to what
those who knew Dr. Darwin well, thought about him. As one of
his granddaughters remarked to me, the term "benevolent" has
been associated with his name, almost in the same manner as that
of "judicious" with the name of the old divine, Hooker. This is
perfectly true, for I have incessantly met with this expression in

letters and in the many published notices about him. To the word
benevolence, sympathy is generally added, and often generosity, as well as hospitality. Mr. Edgeworth says:[46] "I have known
him intimately during thirty-six years, and in that period have
witnessed innumerable instances of his benevolence".

His life-long friend, Mr. Kerr, wrote to my father (May 12, 1802)
about his moral character as follows:

> I think all those who knew him will allow that sympathy and
> benevolence were the most striking features. He felt very sensibly
> for others, and, from his knowledge of human nature, he entered
> into their feelings and sufferings in the different circumstances
> of their constitution, character, health, sickness, and prejudice.
> In benevolence, he thought that almost all virtue consisted. He
> despised the monkish abstinences and the hypocritical pretensions which so often impose on the world. The communication
> of happiness and the relief of misery were by him held as the only
> standard of moral merit. Though he extended his humanity to
> every sentient being, it was not like that of some philosophers, so
> diffused as to be of no effect; but his affection was there warmest
> where it could be of most service to his family and his friends, who
> will long remember the constancy of his attachment and his zeal
> for their welfare.

His neighbour, Sir Brooke Boothby, after the loss of his child
(to whom the beautiful and well-known monument in Ashbourne
church was erected), in an ode addressed to Dr. Darwin, writes in
strong terms about his sympathy and power of consolation.

Trifling circumstances often show a man's character, equally
with greater ones. My father owed a small sum of money to his
father, who asked him (1793) to buy with it a goose-pie (for which
it seems Shrewsbury was then famous) and send it at Christmas
to an old woman living in Birmingham, "for she, as you may
remember, was your nurse, which is the greatest obligation, if
well performed, that can be received from an inferior".

Mr. Edgeworth had [in 1766] corresponded, as a stranger, with my grandfather about the construction of carriages, and came to Lichfield to see him, but found him out. He was asked by Mrs. Darwin to stay to supper:

> When this was nearly finished, a loud rapping at the door announced the Doctor. There was a bustle in the hall, which made Mrs. Darwin get up and go to the door. Upon her exclaiming that they were bringing in a dead man, I went to the hall. I saw some persons, directed by one whom I guessed to be Doctor Darwin, carrying a man who appeared motionless. "He is not dead", said Dr. Darwin, "he is only dead drunk. I found him", continued the Doctor, "nearly suffocated in a ditch; I had him lifted into my carriage, and brought hither, that we might take care of him to-night".

Not many men would have done anything so disagreeable as to bring home a drunken man in their carriage. When a light was brought, the man was found to be, to the astonishment of all present, Mrs. Darwin's brother, "who for the first time in his life", as Mr. Edgeworth was assured, "had been intoxicated in this manner, and who would undoubtedly have perished had it not been for Dr. Darwin's humanity". We must remember that in those good old days it was not thought much of a disgrace to be very drunk.

After the man had been put to bed, Mr. Edgeworth says that he first discussed with Dr. Darwin the construction of carriages and then various literary and scientific subjects, so that "he discovered that I had received the education of a gentleman", "Why, I thought", said the Doctor, "that you were only a coachmaker". "That was the reason", said I, "that you looked surprised at finding me at supper with Mrs. Darwin. But you see, Doctor, how superior in discernment ladies are even to the most learned gentlemen".[47]

There is, perhaps, no safer test of a man's real character than that of his long continued friendship with good and able men. Now Mr. Edgeworth asserts,[48] after mentioning the names of

Keir, Day, Small, Boulton, Watt, Wedgwood, and Darwin, that "their mutual intimacy has never been broken except by death". To these names, those of Edgeworth himself and of the Galtons may be added. The correspondence in my possession shows how true Mr. Edgeworth's statement is with respect to several of them; and they were all men well known at that time, and some of them to the present time.

Mr. Day was a most eccentric character, whose life has been sketched by Miss Seward: he named Dr. Darwin "as one of the three friends from whom he had met with constant kindness";[49] and my grandfather, in a letter to my father, says: "I much lament the death of Mr. Day. The loss of one's friends is one great evil of growing old. He was dear to me by many names (*multis mihi nominibus charus*), as friend, philosopher, scholar, and honest man".

The love of woman is a very different affair from friendship, and my grandfather seems to have been capable of the most ardent love of this kind. From the many MS. verses addressed to his second wife, and about her, before they were married, it is evident that Dr. Darwin was passionately attached to her, even during the lifetime of her husband, who died in 1780; these verses are somewhat less artificial than his published poems.

That Dr. Darwin was charitable, we may believe on Miss Seward's testimony, as it is supported by concurrent evidence. After saying that he would not take fees from the priests and lay-vicars of the Cathedral of Lichfield, she adds: "Diligently, also, did he attend to the health of the poor in the city, and afterwards at Derby, and supplied their necessities of food, and all sort of charitable assistance".[50]

Sir Brooke Boothby also in one of his published sonnets, says:

If bright example more than precept sway,
Go, take your lesson from the life of Day;
Or, Darwin, thine whose ever-open door
Draws, like Bethesda's pool, the suffering poor;
Where some fit cure the wretched all obtain,
Relieved at once from poverty and pain.

The gratitude of the poor to him was shown on two occasions in a strange manner. He was asked to attend one of the Cavendishes at Newmarket during the races—it is believed Lord George Cavendish, his schoolfellow and friend, who was godfather to his daughter Emma (the young lady who was so vehement about Miss Seward's 'Memoirs'), and to whom he bequeathed a legacy of one hundred pounds. My grandfather slept at an hotel, and during the night was awakened by the door being gently opened. A man came to his bedside and thus spoke to him:

> I heard that you were here, but durst not come to speak to you during the day. I have never forgotten your kindess to my mother in her bad illness, and the only way I can show my gratitude is to tell you to bet largely on a certain horse (naming one), and not on the favourite, whom I am to ride, and who we have settled is not to win.

My grandfather afterwards saw in the newspaper that, to the astonishment of everyone, the favourite had not won the race.

The second story is that, as the doctor was riding at night on the road to Nottingham, a man on horseback passed him, to whom my grandfather said 'good night'. As the man soon slackened his pace, my grandfather was forced to pass him, and again spoke to him, but received neither time any answer. A few nights afterwards a traveller was robbed at nearly the same spot by a man who, from the description, appeared to be the same. Here the two editions of the story differ widely; in one it is said that my grandfather out of curiosity visited the robber in prison, who owned that he had intended to rob him, but added: "I thought it was you, and when you spoke I was sure of it. You saved my life many years ago, and nothing could make me rob you."

{These stories appear at first hardly credible, but I have traced them, more or less clearly, through four distinct channels to my grandfather, whose veracity has never been doubted by any one who knew him. The fundamental facts are the same with respect to the jockey story, but the accessories differ to an extreme degree. With respect to the second story even some of the fundamental facts differ, and I feel much doubt about it. It is quite curious how stories get unintentionally altered in the course of years. They were first communicated to me by a daughter of Violetta Darwin, who heard her mother relate them.}

It is remarkable that in so large a town as Derby, and at so late a period as January 1784, there was no public institution for the relief of the poor in sickness. Dr. Darwin therefore at this time drew up a circular, the MS. of which is in my possession, stating that

> as the small-pox has already made great ravages in Derby, showing much malignity even at its commencement; and as it is now three years since it was last epidemic in this town, there is great reason to fear that it will become very fatal in the approaching spring, particularly amongst the poor, who want both the knowledge and the assistance necessary for the preservation of their children.

He accordingly proposed that a society should be formed—the members to subscribe a guinea each—and that a room should be hired as a dispensary, where the medical men of the town might give their attendance gratuitously. The poor were to be directed to take their prescriptions in due order to all the druggists in the town, and it thus appears that opposition was feared from them. The circular then expresses the hope that the dispensary "may prove the foundation-stone of a future infirmary".

If those bigots who now oppose vaccination would consider such incidental notices as those above given of the former prevalence of small-pox, and of its dreadful malignity, they might perhaps distrust their own judgments; but probably they are too ignorant to be able to see their own ignorance. It appears that there are always men with minds so distorted that they will oppose any practice however beneficent, facts however certain, and theories however well established.

In this same year of 1784 he seems to have taken the chief part in founding a Philosophical Society in Derby. The members met for the first time at his house, and he delivered to them a short but striking address, from which the following passages may be given:

> I come now to the second source of our accurate ideas. As we are fashioned and constituted by the niggard hand of Nature with such imperfect and contracted faculties, with so few and such imperfect senses; while the bodies, which surround us, are indued with infinite variety of properties; with attractions, repulsions, gravitations, exhalations, polarities, minuteness, irresistance, &c., which are not cognizable by our dull organs of sense, or not adapted to them; what are we to do? shall we sit down contented with ignorance, and after we have procured our food, sleep away our time like the inhabitants of the woods and pastures? No, certainly!—since there is another way by which we may indirectly become acquainted with those properties of bodies, which escape our senses; and that is by *observing and registering* their effects upon each other. This is the tree of knowledge, whose fruit forbidden to the brute creation has been plucked by the daring hand of *experimental philosophy*.

He concludes the address with the words:

> I hope at some distant time, perhaps not very distant, by our own publications we may add something to the common heap of knowledge; which I prophesy will never cease to accumulate, so long as the human footstep is seen upon the earth.

No man has ever inculcated more persistently and strongly the evil effects of intemperance than did Dr. Darwin; but chiefly on the grounds of ill-health, with its inherited consequence, and this perhaps is the most practical line of attack. It is positively asserted that he diminished to a sensible extent the practice of drinking amongst the gentry of the county.[51]

He himself during many years never touched alcohol under any form; but he was not a bigot on the subject, for in old age he informed my father that he had taken to drink daily two glasses of home-made wine with advantage. Why he chose home-made wine is not obvious; perhaps he fancied that he thus did not depart so widely from his long-continued rule.

Much earlier (Oct. 15, 1772) he wrote to Wedgwood, who was in feeble health: "I would advise you to live as high as your constitution will admit of, in respect to both eating and drinking. This advice could be given to very few people! If you were to weigh yourself once a month you would in a few months learn whether this method was of service to you".

His advocacy of the cause is not yet forgotten, for Dr. Richardson, in his address in 1879 to the "British Medical Temperance Association", remarks: "the illustrious Haller, Boerhaave, Armstrong, and particularly Erasmus Darwin, were earnest in their support of what we now call the principles of temperance".

When a young man he was not always temperate. Miss Seward relates[52] a story, which would not have been worth notice had it not been frequently quoted. My grandfather went on a picnic party in Mr. Sneyd's boat down the Trent, and after luncheon, when (in Miss Seward's elegant language) "if not absolutely intoxicated, his spirits were in a high state of vinous exhilaration", he suddenly got out of the boat, swam ashore in his clothes, and "walked coolly over the meadows towards the town" of Nottingham. He there met an apothecary, whose remonstrances about his wet clothes he answered by saying that the unusual internal stimulus would "counteract the external cold and moisture"; he then mounted on a tub, and harangued the mob in an extremely sensible manner on sanitary arrangements. But it is obvious that this harangue must have been largely the work of Miss Seward's own imagination.

There was, however, some truth in this story, for his widow, who did not believe a word of it, wrote to Mr. Sneyd, whose answer lies before me. He admits that something "similar" did happen, but gives no details, and advises Mrs. Darwin "to take no notice of this part of her {Miss Seward's} very unguarded and scandalous publication".

To show what the gentry of the county thought of her book at the time, I will add that Mr. Sneyd, after alluding in the same letter to her account of the death of Erasmus, remarks: "The authoress deserves to be exposed for her want of veracity and every humane feeling". One of Dr. Darwin's stepsons used always to maintain (as I hear from his daughter) that this half-tipsy freak was due to some of the gentlemen of the party, "who were vexed at his temperate habits", having played him a trick; and this, I presume, means that he was persuaded to drink as something as weak, which was really strong.

He sympathised warmly with Howard's noble work of reforming the state of the prisons throughout Europe, as his lines in the 'Loves of the Plants' (Canto II.) show:

> And now, Philanthropy! thy rays divine
> Dart round the globe from Zembla to the Line;
> O'er each dark prison plays the cheering light,
> Like northern lustres o'er the vault of night.
> From realm to realm, with cross or crescent crown'd,
> Where'er mankind and misery are found,
> O'er burning sands, deep waves, or wilds of snow,
> Thy Howard journeying seeks the house of woe.
> Down many a winding step to dungeons dank,
> Where anguish wails aloud, and fetters clank;
> To caves bestrew'd with many a mouldering bone,
> And cells, whose echoes only learn to groan;
> Where no kind bars a whispering friend disclose,
> No sunbeam enters, and no zephyr blows,
> He treads, inemulous of fame or wealth,

Profuse of toil, and prodigal of health;
With soft assuasive eloquence expands
Power's rigid heart, and opes his clenching hands,
Leads stern-eyed Justice to the dark domains,
If not to sever, to relax the chains. . . .
   The spirits of the Good, who bend from high
Wide o'er these earthly scenes their partial eye,
When first, arrayed in Virtue's purest robe,
They saw her Howard traversing the globe,
Mistook a mortal for an Angel-Guest,
And ask'd what Seraph-foot the earth imprest.
Onward he moves! Disease and Death retire,
And murmuring demons hate him, and admire.

The lines on Slavery already quoted, *and his appeal to the Great God of Justice in 'Phytologia', 1800, p. 77*—['In many plants, sugar is found ready prepared . . . and in the sugar-cane it abounds. . . . Great God of Justice! grant that it may soon be cultivated only by the hands of freedom, and may thence give happiness to the labourer, as well as to the merchant and consumer']—*prove his abhorrence of this great national crime.*

In February 1789, he tells Wedgwood that he had been reading 'Colonel Jack', by De Foe, and suggests that the account there given of the generous spirit of black slaves should be republished in some journal. Again, on April 13th of the same year (1789), he writes:

I have just heard that there are muzzles or gags made at Birmingham for the slaves in our islands. If this be true, and such an instrument could be exhibited by a speaker in the House of Commons, it might have a great effect. Could not one of their long whips or wire tails be also procured and exhibited? But an instrument of torture of our own manufacture would have a greater effect, I dare say.

The dates of his poem and of this letter should be noticed, for let it be remembered that even the slave-trade was not abolished until 1807; and in 1783 the managers of the Society for the Propagation of the Gospel absolutely declined, after a full discussion, to give Christian instruction to their many slaves in Barbadoes.[53]

In some notes made by my father in 1802, it is said that his father's temper was naturally bold, but that a succession of bad accidents made a deep impression on his mind, and rendered him when old very cautious. At the age of about five years he received an accidental blow on the middle of the head from a maid-servant, and ever afterwards a white lock of hair grew there. When a boy fishing with his brothers, they put him into a bag with his feet only out, and being thus blinded he walked into the river, and from his arms being confined, was very nearly drowned. Again, when playing with gunpowder at school, together with Lord George Cavendish, it exploded and he was badly injured. Lastly, through falls, he twice [?] broke the patella of his knee.

His appearance is shown in the Photograph at the beginning of the volume but what Edgeworth says of his pictures not giving a fair idea of his expression should be remembered. He was deeply pitted with the small-pox, but this has been omitted by the painter. His frame was large and bulky, and he grew corpulent when old. Owing to his lameness, he was clumsy in his movements, but my father says that, when young, he was very active.

I have now given as faithful an account as I could of the character of my grandfather. His energy was unbounded. In his day he was esteemed a great poet. As a physician, he was eminent in the noble art of alleviating human suffering. He was in advance of his time in urging sanitary arrangements and in inculcating temperance. He was opposed to any restraint of the insane, excepting as far as was absolutely necessary. He strongly advised a tender system of education. With his prophetic spirit, he anticipated many new and now admitted scientific truths, as well as some mechanical inventions. He seems to have been the first man who urged the use of phosphate of lime in agriculture, which has proved of such great importance to the country. He was highly benevolent, and retained the friendship of many distinguished men during his whole life. He strongly insisted on humanity to the lower animals. He earnestly admired philanthropy, and abhorred slavery. But he was unorthodox; and as soon as the grave closed over him he was grossly and often calumniated. Such was the state of Christian feeling in this country at the beginning of the present century; we may at least hope that nothing of the kind now prevails.

There remains only to be said that Erasmus Darwin died at Breadsall Priory, near Derby, on Sunday morning, April 18th, 1802, in his seventy-first year. A week previously he had been ill for a few days, but had recovered. On the 17th, whilst walking in his garden with a lady, he told her that he did not expect to live long. At night he was as cheerful as usual. On the following morning, the 18th, he rose at six o'clock and wrote a long letter to Mr. Edgeworth,[54] which he did not live to finish, and which contains the following description of the Priory, where he had been living for about three weeks:

> We have a pleasant house, a good garden, ponds full of fish, and a
> pleasing valley somewhat like Shenstone's—deep, umbrageous,
> and with a talkative stream running down it. Our house is near the
> top of the valley, well screened by hills from the east and north,
> and open to the south, where, at four miles distance, we see Derby
> tower.

At about seven o'clock he was seized with a violent shivering fit, and went into the kitchen to warm himself; he returned to his study, lay on the sofa, became faint and cold, was moved into an

Breadsall Priory, where Erasmus Darwin died in 1802, from a drawing by Violetta H. Darwin, 1857.

arm-chair, where without pain or emotion of any kind he expired a little before nine o'clock.

A few years before, he had written to Edgeworth: "When I think of dying, it is always without pain or fear"; but he had often expressed a strong hope that his end might be painless, and so it proved. His medical attendants differed about the cause of his death, but my father did not doubt that it was an affection of the heart. Many years afterwards his widow showed me the sofa and chair, still preserved in the same place, where he had lain and expired. He was buried in Breadsall Church.

ERASMUS DARWIN, M.D., F.R.S.
Born at Elston, near Newark, 12th December, 1731.
Died at the Priory, near Derby, 18th April, 1802.
Of the rare union of Talents
which so eminently distinguished him
as a Physician, a Poet and a Philosopher
His Writings remain
a public and unfading testimony.
His Widow
has erected this Monument
in memory of
the zealous benevolence of his disposition,
the active humanity of his conduct,
and the many private virtues
which adorned his character.

CHARLES DARWIN'S REFERENCES
(indicated by superscript numbers in the text)

1 These statements are taken chiefly from a sketch of his life published by his father, Erasmus, in 1780, together with two of his posthumous medical essays. See also Hutchinson's 'Biographia Medica', 1799,Vol. i., p. 239; also 'Biographie Universelle', Vol. x., 1855; also an article in the 'Gentleman's Magazine', 1801, Vol. lxxi., Pt. ii., p. 604 [C.D. wrongly has September 1st, 1794, Vol. lxiv., p. 794], signed 'A.D.', evidently Professor Andrew Duncan, of Edinburgh.

2 'Harveian Discourse', by Professor A. Duncan, 1824 [pp. 10–12].

3 Author of 'Hereditary Genius', 'English Men of Science', and of other works and papers.

4 Published by one of his descendants in the 'Gentleman's Magazine', Oct. 1808, Vol. lxxviii., Pt. ii., p. 869.

5 I am much indebted to a son of Dr. Sieveking, who brought to England the original letters preserved by the descendants of Reimarus, for permitting me to have them photographed.

6 'Memoirs of the Life of Dr. Darwin', 1804, p. 11–14.

7 J. Cradock, 'Literary Memoirs', 1828, Vol. iv., p. 143.

8 'Memoirs of the Life of Dr Darwin', 1804, p. 125.

9 'The Botanic Garden', part I, canto I, lines 103–114.

10 'Memoirs of R. L. Edgeworth', 2nd ed., 1821, Vol. ii., p. 111. [1st ed., Vol. ii., p. 131]

11 Dated June 23, 1792 and published in the 'Monthly Magazine', 1803, Vol. ii., p. 100.

12 'Pursuits of Literature'. A Satirical Poem in Four Dialogues; 14th ed., 1808, p. 54.

13 'Loves of the Plants', 1789, Interlude between Cantos I and II.

14 'Monthly Magazine or British Register', 1802, Vol. xiii., pp. 457–463.

15 'Memoir of Maria Edgeworth', 1867, Vol. i., p. 31.

16 'Sketch of the Life of James Keir', F.R.S.', p. 111. [ed. J. K. Moilliet, 1868]

17 'History of Philosophy', 3rd ed., 1867, Vol. ii., pp. 356–8.

18 'Zoonomia', Vol. i., p. 27.

19 'Müller's Elements of Physiology', translated by Baly, 1842, p. 943 [refers to 'Zoonomia', section X, Vol. i., pp. 49–53]

20 'Zoonomia', 1794, Vol. i., p. 99. I was led to search for this passage by its having been given by Dr. Dowson in his 'Erasmus Darwin: Philosopher, Poet, and Physician', 1861, p. 46.

21 See, for instance, Dr. Baeta's work, 'Comparative view of the Theories and Practice of Drs. Cullen, Brown, and Darwin', published in 1800.

22 'Zoonomia', Vol. ii., 1796, p. 352.

23 'Pathology of Mind', 1879, p. 229.

24 'Zoonomia', Vol. ii., 1796, p. 570.

25 'The Temple of Nature', 1803, Notes, p. 45; published after his death.

26 'Elements of Physiology', translated by Baly, 1842, p. 1421.

27 I am indebted to Dr. Dowson's 'Life of Erasmus Darwin', for the reference to the 'Life and Works of Sir. J. Sinclair'. [Vol. ii., p. 85]

28 'Temple of Nature', Notes, p. 120; p. 107 on the phonetic alphabet. See also 'Memoirs of R. L. Edgeworth', Vol. ii., p. 178. [1st ed., Vol. ii., p. 198]

29 'Philosophical Transactions', 1785, Part I, p. 1. [Vol. 75]

30 'Memoirs of R. L. Edgeworth', 2nd ed., 1821, Vol. ii., p. 110. [1st ed., Vol. ii., p. 129]

31 'Journals of Dr. Whalley', 1863, Vol. ii., pp. 73, 220–222.

32 'Monthly Magazine', Vol. xiii., 1802, pp. 457–463.

33 J. Cradock, 'Literary Memoirs', 1828, Vol. iv., p. 271.

34 'Temple of Nature', 1803, p. 54, footnote.

35 Campbell's 'Lives of the Lord Chancellors', Vol. v., 1846, p. 656.

36 'Memoirs of the Life of Dr. Darwin', p. 132.

37 'Memoirs of R. L. Edgeworth', 2nd ed., 1821, Vol. ii., p. 245. [1st ed., Vol. ii., p. 268]

38 'Monthly Magazine', Vol. xiii., 1802, pp. 457–463.

39 'Monthly Magazine', Vol. xiv., 1802, pp. 115–116.

40 'Memoirs of R. L. Edgeworth', 2nd ed., Vol. ii., p. 177. [1st ed., Vol. ii., p. 197]

41 'Monthly Magazine', Vol. xiv., 1802, p. 115.

42 'Journals of Dr. Whalley', edited by Wickham; not published until 1863, Vol. i., p. 342.

43 'Life of Mary Anne SchimmelPenninck', 1858 [particularly Vol. i., pp. 151–3 and 262]

44 'Loves of the Plants', 1789, Canto III, p. 131.

45 'Temple of Nature', 1803, p. 124, footnote.

46 'Monthly Magazine', Vol. xiv., 1862, p. 115.

47 'Memoirs of R. L. Edgeworth', 2nd ed., Vol. i., p. 158. [1st ed., Vol. i., pp. 163–5]

48 Ibid., Vol. i., p. 181. [1st ed., Vol. i., p. 185]

49 Ibid., Vol. ii., p. 113. [1st ed., Vol. ii., p. 133]

50 'Memoirs of the Life of Dr. Darwin', 1804, p. 5.

51 The following short history of temperance societies is extracted from Dr. Krause's MS. notes on Dr. Darwin: "The oldest temperance societies were founded in North America in 1808 by the efforts of Dr. Rush, and in Great Britain in 1829, chiefly at the suggestion of Mr. Dunlop. See Samuel Couling, 'History of the Temperance Movement in Great Britain and Ireland, from the earliest date to the present time', London, 1862. In Germany, indeed, the Archduke Frederick of Austria had founded a temperance order as early as 1439, which was followed in 1600 by the temperance order established by the Landgrave of Hesse, but these were only imitations of the Templars and other orders of knighthood, which

sought by vows to suppress the coarse excesses of drinking bouts, as is indicated by the motto of the first-mentioned order: 'Halt Maas!' The suggestion of the establishment in Germany of true temperance societies on the American and English model was due to King Frederick William III''.

52 'Memoirs of the Life of Dr. Darwin', pp. 64–68.

53 W. Lecky, 'History of England [in the Eighteenth Century]', 1878, Vol. ii., p. 17.

54 'Memoirs of R. L. Edgeworth', 2nd ed., Vol. ii., p. 242. [1st ed., Vol. ii., p. 264]

Notes on the Text
by the editor

───────────────── ⁓ ─────────────────

These notes can either be welcomed as agreeably varied or condemned as a mere hotchpotch. My first aim is to provide manuscript or published sources for C.D.'s quotations. Secondly, I identify the people named, giving dates and brief details, with the addition of 'see DNB' for those included in the *Dictionary of National Biography*, and hence also in the new *Oxford DNB* (2004?). Thirdly, I add further factual information when necessary – and it often is, because C.D. gives few details of Erasmus's life. Finally, I offer occasional comments on C.D.'s text.

Throughout the notes I refer to Erasmus Darwin as 'Erasmus'. His grandson Charles Darwin, I continue to call 'C.D.' when he is in his role as an author; otherwise I call him 'Charles'. I refer to Erasmus Darwin's second son, Erasmus, as 'Erasmus junior'. I identify Erasmus's eldest son, Charles (after whom C.D. was named), as 'the elder Charles' or 'Erasmus's son Charles'.

For details of Erasmus's life, I usually refer to the appropriate pages in my recent biography, *Erasmus Darwin: a Life of Unequalled Achievement* (de la Mare, London, 1999), which I abbreviate as '1999 *Life*'. References to manuscripts and specialized papers can be found there.

Other abbreviations in the notes are as follows:

> DAR, followed by a number, indicates relevant manuscript material in the Darwin Archive at Cambridge University Library.
> DSB = *Dictionary of Scientific Biography*, 16 volumes. Scribners, 1970–80.

Letters = The Letters of Erasmus Darwin, ed. D. King-Hele.
Cambridge University Press, 1981. (A second edition is in
preparation.)

Pearson, Galton = K. Pearson, The Life, Letters and Labours of
Francis Galton, 4 vols. Cambridge University Press, 1914–30.

Pedigrees = R. B. Freeman, Darwin Pedigrees. London, 1984.

UCL = University College London.

Page and line numbers are given to define the subject of each note.

p.5, general   The second and third paragraphs of the Preface are
useful, but the other three are rather confused.
Krause's 28-page original article in Kosmos, pages
397–424, was entitled 'Erasmus Darwin, the
grandfather and forerunner of Charles Darwin'. In the
first paragraph here, C.D. implies that this article has
been translated, though he gives it a different title. In
reality, as mentioned in the fourth paragraph, Krause
'added largely to his essay', and it is this much longer
version which was translated.

The book as published in November 1879 consisted
of C.D.'s 129-page 'Preliminary Notice' and Krause's
86-page essay, which had yet a third title, 'The Scientific
Works of Erasmus Darwin'. However, 26 further pages
of the translation of Krause's essay (DAR 227.8:50)
were omitted; they reviewed past ideas of Earth history.

p.5, line 2   Dr Ernst Ludwig Krause (1839–1903) was a German
botanist, science writer and one of the editors of
Kosmos. See DSB.

p.5, line 7   'Mr Dallas' is William S. Dallas (1824–1900), Assistant
Secretary of the Geological Society, and author of
Natural History of the Animal Kingdom (1856) and Elements
of Entomology (1857).

p.5, line 13   Erasmus's Commonplace Book passed from his widow
Elizabeth, on her death in 1832, to her only living son
Sir Francis Darwin (1786–1859), and on his death to his
son Reginald Darwin (1818–1892), who lent it to

Charles. It is now owned by English Heritage (Down House) and is (in 2002) on loan to Erasmus Darwin's House at Lichfield, where a complete photocopy can be consulted. There are also microfilms at Cambridge University Library and Derby Local Studies Library.

p.5, line 21   'Miss Seward' is the poet Anna Seward (1742–1809), who figures prominently later (see notes to pp.26 and 70). 'Dr Dowson' is John Dowson (1805–1879), who graduated M.D. at Glasgow in 1829, wrote *An Introduction to . . . Medicine* (1834) and was for many years a medical practitioner at Whitby, where he gave the lecture in 1860.

p.6, line 10   Samuel Butler (1835–1902) was author of *Erewhon* (1872) and of several books insisting on the role of mind in evolution. See DNB. By a most unfortunate chance, Butler's book *Evolution: Old and New* was published while Krause was enlarging his essay. Butler attacked C.D.'s theory of evolution, and Krause added some oblique criticisms of Butler, though without mentioning his name.

   In January 1880 Butler accused Charles of sharp practice, because the published version of the Preface did not mention that Krause had enlarged his essay, and readers would think that the oblique criticisms were in the original German version. This niggling complaint upset Charles, who had been careful to mention the enlargement in the fourth paragraph (line 4 here), which was later deleted. A full account of the episode and its aftermath is given by Nora Barlow in *The Autobiography of Charles Darwin*, ed. N. Barlow, Collins, 1958, pp.167–219. For relevant manuscript material, see DAR 92:B65, 92:B108, 133.5:2, 143:23, 143:421–7.

p.7, line 1   For some of the original wording of this paragraph, see the manuscript first draft in DAR 212.1. The paragraph was deleted by Henrietta and is therefore in italics.

p.7, line 9   The book mentioned (which is not really 'large') is part of British Library MS Cotton Vespasian E.16, and extends to 22 handwritten pages. It is a continuation, by William Whitlock (d.1584), of an existing Chronicle of the Bishops of Lichfield up to 1347 by Thomas Chesterfield (d.1451–2). Whitlock extends the Chronicle to 1559. For Chesterfield, Whitlock and Sir Robert Cotton, see DNB.

　　　　　　Charles's brilliant son George unearthed this evidence of an early scholarly Darwin who knew Cotton (see DAR 210.14:26): his discovery gives the book a wonderful start. And as a bonus there is the link with Lichfield. Cotton's writing is still legible enough to allow a few minor corrections to C.D.'s quote.

p.7, line 15   Here the two lines beginning 'of yeomen who . . .' are taken from C.D.'s original manuscript (DAR 133.5:1). The rest of the paragraph is from the first proof. For further details of the ancestral Darwins, see *Pedigrees* and the family tree in Appendix B.

p.7, line 23   For the locations of Marton and Cleatham, see the map on p. 4.

p.8, line 19   For Erasmus Earle (1590–1667), see DNB; and Pearson, *Galton*, Plate LXV.

p.8, line 23   C.D. has 'Wilsford' instead of 'Wilford', but I believe this is an error. See Pearson, *Galton*, p.10, and *Pedigrees*, p.6. For the locations of Wilford and Elston, see the map on p. 4.

p.10, line 3   C.D. 'supposes' that the Elston estate was bequeathed to Robert, and I accept this view in the 1999 *Life*. But his mother's estate may have passed to Robert's elder brother the fourth William on her death in 1722; if so, William allowed Robert to occupy Elston Hall, which their mother had owned and lived in. (Their father, the third William, had died in 1682 when Robert was only 16 days old.) When the fourth William died in 1760, his estate passed to his only living son Dr William Morgan Darwin (1710–1762): as the latter had no male offspring, it may be that his nephew Robert Waring Darwin (Erasmus's eldest brother) inherited Elston Hall from him in 1762 (and not from his father in 1754). This alternative scenario was suggested in a letter to Charles in April 1879 (DAR 99.138) from Charlotte Darwin (1827–1885), whose father owned Elston Hall from 1816 till 1841. See family tree, Appendix B.

p.10, line 11   The skeleton was the first to be recorded of a plesiosaur: see 1999 *Life*, pp.2–5, for a picture of it and further details. Dr William Stukeley (1687–1765) was a well known antiquarian and a friend of Isaac Newton. See DNB. This was a major discovery by Robert Darwin, whom Stukeley called 'a person of curiosity'. Stukeley was correct when he said that 'the like whereof has not been observed before in this island'. A complete plesiosaur was not found until Mary Anning's discoveries at Lyme Regis a hundred years later. If the Royal Society had been able to foresee this, Robert might have been elected a Fellow. (He did attend a meeting, presided over by Newton.)

It is rather strange that Charles does not give his great-grandfather any credit for his part in this significant evolutionary discovery. Charles did mention it 40 years earlier in a letter to his second cousin William Darwin Fox (1805–1880), and jokingly said 'that *we* have a right of hereditary descent to be naturalists and especially geologists'– *The Correspondence of Charles Darwin*, Vol.2, p.235 (1986). The fossil still exists, and is on display at the Natural History Museum in London.

p.10, line 19   Dr Thomas Okes (1730–1797), a medical student friend of Erasmus at Cambridge, acted as an army surgeon in the Seven Years War, and later practised as a physician in Exeter. There is a memorial to him in the Cathedral.

p.11, line 1   Apart from quoting the rhyme, C.D. says nothing about Robert's wife (Erasmus's mother). She was born as Elizabeth Hill, was brought up at Sleaford, married Robert Darwin on 1 January 1724 at the age of 21, and then had seven children in eight years. They were all healthy, and lived to ages between 57 and 92. She continued to live at Elston Hall after her husband's death in 1754, and died at the age of 94 in 1797. Erasmus said of her, 'a better mother never existed'.

p.11, line 3   Erasmus's eldest brother Robert Waring Darwin (1724–1816) was a lawyer at Lincoln's Inn before retiring in his thirties to live at Elston Hall as squire and owner. He and Erasmus remained close, and shared an interest in botany and gardening. Robert's *Principia Botanica*, an introduction to Linnaean botany, was published in 1787, with a third edition in 1810. For further details, see 1999 *Life*, pp.216–19, and R. F. Scott, *Admissions to the College of St John the Evangelist in the University of Cambridge*, Part III (1903), pp.538–9.

p.11, line 19   William Alvey Darwin (1726–1783) became a lawyer in London, at Gray's Inn, married Jane Brown in 1772 and rarely met Erasmus. Ownership of Elston Hall passed to his son William Brown Darwin (1774–1841) in 1816.

p.11, line 21   The third brother John Darwin (1730–1805) was only one year older than Erasmus, and they went to school and university together. They remained on good terms. John was rector of Elston, and of Carlton Scroop 20 miles to the east, from 1766 until 1805. He seems to have been conscientious and religious: some of his letters are mini-sermons. For further details, see Scott, *Admissions* . . . (cited above), p.601.

p.11, line 27   In what seems like a post-modern tease, Charles decides to tell us first about Erasmus's children, rather than Erasmus himself. Though baffling initially, this ruse can be seen to spring from his strong belief in the role of heredity, the *sine qua non* of evolution. To him there was a witty logic in presenting the past and future before the present. However, Henrietta did not agree: she moved the life stories of the children to nearer the end of the book.

p.11, line 31   The information about Erasmus's son Charles is largely from the book *Experiments establishing a Criterion between Mucaginous and Purulent Matter* by C. Darwin (Cadell, 1780). The book, compiled by Erasmus, included the prize essay and a short biography of Charles (pp.127–35). For details of Charles's visit to France, see 1999 *Life*, pp.72–5.

p.12, line 14   I have replaced 'was sent' by 'went', because Charles entered Christ Church, Oxford (in March 1775) against the wishes of his father. Presumably it was his teenage rebellion.

p.12, line 28   Dr Andrew Duncan (1744–1828) was a man of great kindness and energy who became one of the leading figures in Scottish medicine (see DNB). In his Harveian Discourse of 1824, p.10, Duncan wrote of the elder Charles: 'During his studies he exhibited such uncommon proofs of genius and industry as could not fail to gain the esteem and affection of every discerning teacher'. A year later Duncan spoke with 'the warmest affection' about the elder Charles to the younger Charles. This extraordinary link becomes all the more poignant if you credit the possibility that the younger Charles's life-work on evolution might have been pre-empted by his brilliant uncle, if he had lived.

p.13, line 5   Erasmus's letter from Edinburgh (DAR 227.1:46) was written to his second son Erasmus junior. The death of his son Charles was the worst event in Erasmus's life, because of the immense talent and promise unfulfilled. Charles's life was celebrated in an Elegy,

most of which was probably written by Erasmus. See
1999 *Life*, pp.142–5.

p.13, line 6 Josiah Wedgwood (1730–1795) scarcely needs
identifying. C.D. gives no hint that this is his other
grandfather!

p.13, line 11 The original first draft of C.D.'s account of Erasmus
junior is available: DAR 212:3.

p.13, line 23 'Boulton' is the great manufacturer Matthew Boulton
(1728–1809), and 'Day' is Thomas Day (1748–1789),
who will appear more directly later (see p.124). Both
were lifelong friends of Erasmus senior.

p.13, line 30 Erasmus junior left Lichfield when he was 22, and set
up a law practice in Derby. Hence my correction of the
text.

p.14, line 4 The first half of the original first draft of C.D.'s
account of his father Robert is available: DAR 212:4
and 5. Charles wrote at greater length about his father
in his *Autobiography*, pp.28–42.

p.14, line 9 Wheatstone is Sir Charles Wheatstone (1802–1875),
famous for his researches in acoustics, optics and
electricity. See DSB.

p.14, line 24 Robert went to Shrewsbury in 1786 with his brother
Erasmus and Archdeacon Robert Clive (1723–1792), a
friend of the family who lived nearby. The first payment
of £20 to Robert from his father, plus the £20 from his
uncle John, would be equivalent to nearly £4000 today.
It is still surprising that Robert never needed more,
though the fact that his father was the most famous
doctor of the region would have been helpful.

p.15, line 24 The story of this loan also figures in C.D.'s
Autobiography, p.29, where the manufacturer appears as
'Mr B—', possibly Charles Bage, son of Erasmus's

friend Robert Bage. The loan of £10,000 is equivalent to about £¹/₂ million today, and the story reflects Robert's great success as a financier, made possible initially by his marriage to Susannah Wedgwood, who received £30,000 under the will of her father Josiah Wedgwood I (1730 1795). All Robert's children were left quite rich as a result.

p.16, line 12   Members of the family have now formed an overlapping succession of Fellows of the Royal Society for six successive generations. See *Notes Rec. Roy. Soc.*, Vol.52, p.159 (1998).

p.16, line 14   For the children of Erasmus's second wife, see 1999 *Life*, pp.355–6 and the family tree in Appendix B. The cavalry officer was Edward (1782–1829); the rector of Elston was John (1787–1818). 'Captain Darwin', the son of Francis, was Edward Levett Darwin (1821–1901). Erasmus's daughter Violetta (1783–1874) is the source for several stories told by C.D. The other two daughters were Emma (1784–1818), beautiful but plagued by illness; and Harriot (1795–1825), who married Captain (later Admiral) T. Maling.

p.16, line 29   Erasmus's interest in mechanics was apparently innate, but it was his eldest brother Robert (1724–1816) who led him into poetry.

p.17, line 11   The spelling in the original manuscript (Commonplace Book, p.201) is worse than in this printed text. (In verse 4, 'pouched' may be interpreted as 'poached', but looks like 'ponshed'.) The poem shows Erasmus's 'country-sports' background, which he rejected in later years.

p.18, line 1   Susannah (1729–1789) was Erasmus's favourite sister. She acted as housekeeper for him in 1757 and from 1770 to 1781. Her 'Diary in Lent' is dated 1748, because in the old-style calendar the new year-number (here 1749) was not used until 25 March. The ages of Susannah and Erasmus given by C.D. therefore need to be increased: Susannah was 19, Erasmus 17.

p.18, line 12   The manuscripts of Susannah's letter and 'Diary in Lent', (both in DAR 227.3:1) have been interpreted well by C.D.'s transcriber. Here and in Erasmus's reply, I have restored abbreviated words, such as y^e and y^n. See *Letters*, p.xvii.

p.19, line 28   Erasmus's letter to Susannah is based on his draft manuscript, now DAR 227.1:1. For comments on the letter, see *Letters*, p.4. There is also (DAR 267) a manuscript copy of the letter as sent, made by Erasmus's father; this was not available to C.D., and so I have not used it.

p.21, line 1   C.D.'s note about the letter becoming 'hardly legible' is hardly accurate. It reads (in part): ". . . but I can't help thinking that was you to dowse pretty well the fore-cited divine, he would make fish far superior to this Hogg, and what food can be properer in Lent''. Evidently this schoolboy banter was thought too improper for publication.

p.21, line 4   The first letter (28 December 1749) was to his school-friend Samuel Pegge (1733–1800), later a musical composer, poet, antiquary, barrister and historian. See DNB, and *Letters*, pp.4–6. The second letter (11 December 1750) was to William Burrow (1683–1758), headmaster of Chesterfield School from 1722 to 1752: under his guidance the school gained a high reputation. For Burrow, see *Letters*, pp.6–7.

p.21, line 12   Erasmus's eldest brother Robert was admitted to St John's College seven years earlier, on 1 July 1743, but he never took a Cambridge degree, having pursued a legal career at Lincoln's Inn, where he was admitted in June 1743.

p.21, line 25   Erasmus may have studied mathematics, as C.D. says, but there is no record of his taking a Bachelor of Arts degree. He took his M.B. in 1755.

p.21, line 30   The poem on Prince Frederick was first published in 1751. See 1999 *Life*, p.11.

p.22, line 1   Erasmus set out for Edinburgh with his brother Robert on 1 October 1753. They arrived on 10 October. The receipt mentioned is in the Commonplace Book, p.202.

p.22, line 6   For James Keir, see p.114. For the manuscript of his illuminating letter, see DAR 227.6:76. Dr Robert Whytt (1714–1766) lectured on the theory of medicine. See DNB. Hermann Boerhaave (1668–1738), Professor at Leyden, wrote standard works on medicine and chemistry, and much influenced teaching and progress in both subjects. Professor William Cullen (1710–1790) was the most famous of the medical teachers at Edinburgh. See DNB.

p.23, line 5   For the full text of this letter, notes on it and a translation of the Latin, see *Letters*, pp.8–9. Okes has already been identified in my notes for page 10.

p.24, line 1   Erasmus did not succeed in starting a practice at Nottingham, as C.D. correctly says. But he did have one patient, 'shoemaker B', who was stabbed by another shoemaker, 'A', in a drunken fracas. Unfortunately the patient died, leaving his doctor with no kudos and no fee. See 1999 *Life*, p.20.

p.24, line 3   The friend was Dr Albert Reimarus (1729–1814), son of Hermann Reimarus (1694–1768), whose deistic philosophy had some influence on Erasmus, and on Keir. Albert Reimarus later became a physician in Hamburg. See *Letters*, p.10, and 1999 *Life*, pp.17 and 86–7.

p.24, line 11   'Dr Hill' is John Hill (1714–1775), a prolific and controversial author. See DNB. Mr G—y is

Thomas Gurney (1705–1770), clockmaker and pioneer shorthand writer, whose system was used in Parliament until 1914. See DNB. By chance, Gurney is mentioned again in the next paragraph. In the manuscripts of the letter to Reimarus (available as DAR 227.1:2–7), Mr D. is given as Mr Douglass, but I have not identified him.

p.26, line 2   For details of Erasmus's early life at Lichfield, see 1999 *Life*, Chapters 2–4. 'All his works' were published after he left Lichfield; so C.D.'s remark is rather off-target.

p.26, line 12   Anna Seward (1742–1809), later known as the 'Swan of Lichfield' and perhaps the leading English woman poet of the 1780s, wrote *Memoirs of the Life of Dr Darwin* (1804), which gives unique but not always accurate details of Erasmus's early life at Lichfield. C.D. is critical of her inaccuracy, here and later in the book. To be fair to Anna, she does claim to be reporting what Polly told her.

p.26, line 22   This letter, with notes, appears in *Letters*, pp.12–15. C.D.'s text may be from the original manuscript. See DAR 227.1:15 for a nineteenth century copy.

p.27, line 30   Erasmus was staying with the Jervis family at Darlaston manor house, near Stone. 'Mrs Jervis' was probably the aunt of John Jervis, later Admiral and Earl of St Vincent (1735–1823).

p.28, line 1   'Mr Howard' is Polly's father Charles Howard (1707–1771), a proctor in the Ecclesiastical Court at Lichfield. He was a schoolfellow and friend of Samuel Johnson (1709–1784), who called him 'a cool and wise man'. Howard's wife Penelope had died in childbirth when Polly was eight years old. See 1999 *Life*, pp.31–2. Howard brought the name 'Charles' into the Darwin family.

p.28, line 10   'My sister' is Susannah, who was acting as housekeeper for Erasmus in 1757.

p.28, line 14 Nelly White (c.1742–1823) was known as the 'Belle of Lichfield' at the time of her marriage to Revd. Fyge Jauncey in 1767.

p.29, line 10 The value of money provokes endless argument. My own choice for converting money amounts in the years 1750–1780 to the equivalent in 2002 is to multiply by about 100. Thus Erasmus's earnings of £192 for the year 1757 would be equivalent to about £19,000 today, and his 1772 earnings of £1025 would be equivalent to about £100,000 today. Between 1780 and 1800 the value of money declined by about 40%, and a lower multiplying factor is appropriate. See 1999 *Life*, p.27 and the references in note 8 of Chapter 2.

p.30, line 1 The quotation is from Erasmus's letter to Robert of 10 May 1799 (DAR 227.1:162).

p.30, line 12 Erasmus gave free treatment to the local poor, and his income depended on long journeys to the distant rich, quite often to inoculate children against smallpox. Inoculation, brought in by Lady Mary Wortley Montagu and Princess Caroline in the 1720s, was one of the few success stories in English medicine of the eighteenth century. It reduced smallpox mortality from perhaps 15% to possibly as low as 2%. See P. Razzell, *The Conquest of Smallpox* (1977) and I. Grundy, *Lady Mary Wortley Montagu* (1999).

p.30, line 17 For the Commonplace Book, see my notes on the preface (p.98).

p.30, line 23 Joseph Cradock (1742–1826) was a country gentleman with a taste for literature. See DNB.

p.30, line 25 Erasmus was very hesitant about publishing his poetry because he thought it would damage his medical practice. The poem he sent to Cradock is the first of his love-poems to Elizabeth Pole, who is the 'Derbyshire lady' he mentions. For the letter and the poem, see *Letters*, pp.75–8.

p.31, line 1   C.D.'s mistake over the name of Colonel Pole (1717–1780) has been copied in dozens of reference books and scholarly studies. The Colonel's name was Edward Sacheverel Pole (though the spelling 'Sacheverell' is sometimes seen). He inherited Radburn Hall in 1765 from his uncle German Pole, who had it built. Colonel Pole survived arduous service in the Seven Years War (1756–63), being left for dead on the field in three battles; during the battle of Minden a bullet entered his left eye and came out at the back of his head. He probably met Elizabeth at nearby Kedleston Hall, the palatial home of her half-sister Caroline Lady Scarsdale (previously Lady Caroline Colyear).

     Elizabeth (1747–1832) was the illegitimate daughter of the second Earl of Portmore (1700–1785). She married Colonel Pole in 1769. Erasmus was the Poles' doctor by 1771, and fell in love with Elizabeth by 1775; after that he kept up a wooing-in-verse until the Colonel's death in 1780, and married Elizabeth in March 1781. For further details, see 1999 *Life*, especially pp.125–8, 138–40 and 170–5.

     The elder son of Colonel Pole and Elizabeth, namely Sacheverel Pole (1769–1813), changed his surname in 1807 by royal charter to 'Chandos-Pole', after proving his descent from a sister of the warrior knight Sir John Chandos, who died in 1370. C.D. knew of the Victorian Chandos-Poles, but was unaware of the name-change, which occurred after Erasmus's death.

p.31, line 8   'There is little to relate of his life . . .' This one-liner is among the best of Charles's jokes: 'I chuckled much', he says in one of his letters, and he may well have done so here. I presume he suddenly realised that his essay would be far too long if he went on quoting the letters in full – he probably had about 250 letters available. So, with this witty escape-clause, he slides out of any further narrative of the last 45 years of Erasmus's life. For more about these 45 years, see 1999 *Life*, pp.25–345.

p.31, line 11 The site of the botanic garden is still rural today, though only 1½ miles from the centre of Lichfield. In recent years it has been a deer-park, but the lakes created by Erasmus may still be seen in wet weather. The 'Handbook for Lichfield' is *Handbook for the City of Lichfield* by John Hewitt (Lichfield, 1874), p.35.

p.31, line 24 *The Loves of the Plants* was officially published in April 1789, but some copies were available in the autumn of 1788, as C.D. says. The 'letter to my father' is DAR 227.1:98.

p.31, line 31 The success of *The Botanic Garden* was indeed immediate. Erasmus was acclaimed as the leading English poet of the 1790s: see 1999 *Life*, pp.264–5. Charles realized this, but was diffident about saying so: he did not wish to blow the family trumpet too loudly.

p.32, line 1 There were many other later editions and translations of *The Botanic Garden*. See 1999 *Life*, pp.401–2. Erasmus himself says (*Letters*, p.225) that he received £900 for *The Botanic Garden*: he was probably referring to the 'part which appeared the second' (*The Economy of Vegetation*).

p.32, line 8 Horace Walpole (1717–1797) was probably the most influential of current literary pundits, the 'Prime Minister of Taste' as he has been called. His enthusiasm for *The Botanic Garden* helped in its success. Erasmus's confident portrayal of the origin of the Universe put him in a class of his own among contemporary poets. See 1999 *Life*, pp.258–9.

p.33, line 1 Richard Lovell Edgeworth (1744–1817) was an active inventor in his early years. He admired Erasmus's method of improved steering for carriages, met him in 1766, and became a close friend – and member of the 'Lunar' group. Although he lived mainly in Ireland, Edgeworth kept in touch by correspondence and, after Erasmus died, defended him against detractors. 'The Ballet of Medea' is in *The Loves of the Plants*, Canto III, lines 135–178. For Edgeworth, see D. Clarke, *The Ingenious Mr Edgeworth* (1965).

p.33, line 19   William Cowper (1731–1800) and William Hayley (1745–1820) were among the best-known English poets of the 1780s, and their wholehearted approval of Darwin's poem also enhanced its reputation.

p.34, line 3   The *Pursuits of Literature* (1794) was a rather shallow satirical hatchet-job on all current poets, published anonymously but now known to be by T. J. Mathias (1754?–1835). See DNB. Erasmus, though only lightly criticized, responded appropriately:

> You've sure the poetaster's curse,
> Who blame bad poetry in worse.
> (UCL Pearson Papers 577, p.48)

p.34, line 8   'The Loves of the Triangles' was a 294-line poem published in April–May 1798 in three issues of the *Anti-Jacobin*, a periodical created to combat subversive ideas and to boost morale in the war against the French. The originator of the *Anti-Jacobin*, and co-author of the poem, was George Canning (1770–1827), Under-Secretary for Foreign Affairs in Pitt's government and a future Prime Minister. The poem is an amusing parody of *The Loves of the Plants*, but the bite is in the notes, which ridicule three of Erasmus's ideas: that electricity will have important future uses; that human beings have evolved from lower forms of life; and that mountains are much older than the orthodox 6000 years.

p.34, line 11   Wordsworth and Coleridge attacked the 'gaudiness and inane phraseology' of Erasmus's verse, in the Advertisement to *Lyrical Ballads* (1798).

p.34, line 22   Despite his previous hesitations, C.D. does acclaim Erasmus as 'a great master of language'.

p.34, line 26   C.D. is quoting from *English Bards and Scotch Reviewers*, where Byron attacks nearly all poets, living or freshly dead. Lines 891–902 are on Erasmus.

p.34, line 28   Erasmus's first paper in the *Philosophical Transactions* was in fact of considerable value. He disproved by an elegant experiment the widely-held view that vapours ascend only if electrified. The paper was a step towards Erasmus's major achievement in formulating the law of adiabatic expansion of gases, which he then applied to explain how most clouds form. (Phil. Trans., Vol. 78, pp.43–52, 1788). As C.D. was not an atmospheric physicist (and I was!), I am surprised that he dared to condemn the paper as 'of no value'.

p.35, line 14   Jesse Foot (1744–1826) was a surgeon who disliked John Hunter (1728–1783) and pursued him with pamphlets. See DNB for Foot, and DSB for Hunter.

p.35, line 16   C.D.'s witticism about *Zoonomia* being 'honoured by the Pope' has a double sting, because Erasmus's crime was his advocacy of biological evolution. The many editions of *Zoonomia*, including three Irish and four American, as well as the translations, are given in 1999 *Life*, p.402.

p.35, line 20   'I need say little on this head'. Another joke. There are six more pages to come.

p.35, line 22   Here C.D. almost admits he is wrong in his prejudice that Erasmus only made speculations, not experiments. Maria Edgeworth (1768–1849), well known as a novelist, was the eldest daughter of R. L. Edgeworth. See M. Butler, *Maria Edgeworth* (1972).

p.36, line 8 James Keir (1735–1820) was Erasmus's earliest lifelong friend: they met at Edinburgh as students. Keir, as well as being 'a mighty chemist' and a pioneer of the chemical industry, was a very judicious man, who often acted as chairman at Lunar Society meetings. When Erasmus hesitated over publishing *The Botanic Garden*, he asked for Keir's advice and took it. Similarly with *Zoonomia* – Keir advised splitting it into two volumes. On chemistry, however, as C.D. says, Erasmus led while Keir stuck to the old ideas; his throwaway term 'oxyde hydro-carbonneux' is, ironically, much the same as 'carbohydrate'.

p.36, line 16 The poor opinion of Erasmus's views on psychology held by Lewes and C.D. is not so widely shared today: see E. S. Reed, *From Soul to Mind* (1997) and R. Porter, *Enlightenment* (2000), pp.435–445.

p.36, line 32 Erasmus's 'law' of associated movements is discussed in *Zoonomia* i 49–53, and Müller's comments are on p.943 of his book, as indicated in C.D.'s note 19. For 'the illustrious Johannes Müller' (1801–1859), see DSB.

p.37, line 23 Dr Richard Warren (1731–1797), senior royal physician, was singularly useless during King George III's 'madness' in 1788–9, but thereafter enjoyed a hugely lucrative London practice. See DNB.

p.37, line 30 One of the three surgeons was Sir Astley Cooper (1768–1841). The 'daughter of Dr Darwin' was Violetta (1783–1874), mentioned earlier.

p.38, line 15 Sir Robert Christison (1797–1882) was an eminent medical man. See DNB.

p.38, line 27 Henry Maudsley (1835–1918) was a physiologist and psychologist. The third edition of his *Pathology of Mind* appeared in 1879. The quotation from *Zoonomia* is from Volume i, p.25.

p.39, line 6   Dr Lauder Brunton, later Sir Thomas Lauder Brunton
                 (1844–1916) was another leading medical man, who
                 pioneered the use of amyl nitrite in treating angina,
                 and wrote the textbook *Pharmacology and Therapeutics*
                 (1865). See DNB and DSB.

p.39, line 9   'Rosenthal' is presumably Isidor Rosenthal
                 (1836–1915), Professor of physiology at Berlin and
                 author of a book on 'General Physiology of muscles
                 and nerves', which was translated into English in 1881.

p.39, line 25   Francis Galton (1822–1911) was Charles's cousin, being
                 the son of Erasmus's daughter Violetta. He was a
                 pioneer in meteorology and fingerprinting, but is best
                 known for stressing the importance of heredity, and
                 for coining the word 'eugenics'. Pearson, *Galton* is the
                 fullest biography of Galton.

                 Ironically, Galton was not aware of the warning
                 against heiresses by his grandfather Erasmus, and in
                 his book *Hereditary Genius* he reached a similar but
                 weaker conclusion. Had he realised his ignorance, he
                 could have claimed his own reproduction of the idea as
                 a further example of hereditary power.

p.40, line 1   In his last long poem, *The Temple of Nature; or, The Origin
                 of Society*, published in 1803 after his death, Erasmus
                 traces the development of life from its origin as a
                 'single living filament', a microscopic speck in
                 primeval seas, through fishes and amphibians to land
                 animals and eventually 'humankind'. Erasmus's
                 speculative evolutionary picture of life's development
                 was a century ahead of its time. It seems extraordinary
                 that C.D. makes comments on several of the prose
                 notes to the poem, but ignores the evolutionary
                 scenario; presumably it was too speculative for him to
                 endorse or even discuss, though it has proved sound.

                 Strangely enough, however, C.D. did (perhaps
                 without realising it?) adopt Erasmus's radical idea that
                 all life originated from a 'single living filament'

(or 'descended from one primordial form', as C.D. phrases it in the first edition of *On the Origin of Species*, p.484). In later editions, having realised that evolution by natural selection would proceed just as well with several 'parents', C.D. mentions the possibility of 4 or 5 progenitors. The 'single living filament', the most daring of Erasmus's assumptions, has been validated by the results of the human genome project.

p.40, line 16   *Phytologia; or the Philosophy of Agriculture and Gardening*, a massive tome of 612 quarto pages, published in 1800, is the best of Erasmus's books in prose. He avoids the errors that plague the treatments of disease offered in *Zoonomia*. In contrast, *Phytologia* bristles with useful, fruitful or prophetic ideas, including two major advances – the first full formulation of photosynthesis and the identification of the main nutrient chemical elements needed by plants. See 1999 *Life*, p.333–8.

p.41, line 5   Sir John Sinclair (1754–1835) had asked Erasmus to write *Phytologia*: the book was dedicated to him. For Sinclair, see DNB and *The Life and Works of J. Sinclair* (2 vols., 1837).

p.41, line 7   'Hunter' is Dr Alexander Hunter (1729–1809), physician at York and author of *Georgical Essays* (1770–2): the comments on bone-dust were said to be in Vol.2, p.93, but I have not found them.

p.41, line 31   C.D. seems embarrassed by Erasmus's principle of organic happiness, presumably because he felt it was metaphysical hijacking of evolution. However, his own conclusion was just as metaphysical: 'And as natural selection works solely by and for the good of each being, all corporeal and mental endowments will tend to progress towards perfection'. (*Origin of Species*, 1st ed., p.489.)

p.42, line 3   Antoine de Jussieu (1748–1836) put forward his 'noble superstructure' in his *Genera Plantarum* of 1789.

Erasmus was much occupied with plant taxonomy in the years 1780–87, when he was translating Linnaeus; and *The Loves of the Plants* grew out of this work. But he seems to have ignored Jussieu, as Charles says. (The quotation from *Phytologia* is on page 578, at the end of the chapter beginning on page 564.)

p.42, line 12   The 128-page book on *Female Education* is the most homely and readable of Erasmus's books. He wrote it for his illegitimate daughters Susanna and Mary Parker, for whom in 1794 he bought a house at Ashbourne, where they ran a successful girls' boarding school for more than 30 years. Many affluent Midland families sent their daughters to the school, and so the book was quite influential. Erasmus argued strongly against the conventional idea that girls were meant to be weak and empty-headed, and went a long way towards modern norms of sexual equality. See 1999 *Life*, pp.281–4. The three passages quoted by C.D. are from pages 89–90, 48 and 19 respectively. The German translation, by C. W. Hufeland, was published in Leipzig in 1822. There were also Irish and American editions in 1798.

p.43, line 21   Charles's witticism about funambulation is fair comment. Even today it is not among the Olympic sports. The quotation is from p.70 of *Female Education*.

p.43, line 28   The quotations about sanitary reform are from pp.242–3 of *Phytologia*.

p.44, line 3   This review of Erasmus's inventions is one of the most admirable features of the book. Mechanical invention was not among Charles's favourite activities, and he could easily have ignored the subject. Instead he describes or mentions 16 inventions. For the horizontal windmill, see 1999 *Life*, pp.80–1, 85, 157–60 and Fig.11.

p.44, line 13   The 'manifold writer' is Erasmus's superb copying machine, with which the first known 'perfect' copy of a document was made in 1779: see 1999 *Life*, pp.151–5 and Plate 9. Erasmus constructed a new machine for his son Robert in 1800 (see 1999 *Life*, p.331), but I have not seen any further mention of it: when it ceased to work well, it was presumably thrown away because no-one knew how to put it right.

The other major invention to be fully engineered by Erasmus (but not known to Charles) was his improved method of carriage steering, devised about 1759 and road-tested successfully over 20,000 miles on his own carriages. The technique was later re-invented and used in nearly all early modern cars. See *Notes and Records of the Royal Society*, Vol.56, pp.41–62, 2002.

Nearly all the other inventions mentioned by Charles are from sketches in the Commonplace Book.

p.44, line 28   Sir Joseph Whitworth (1803–1887) was an eminent mechanical engineer and manufacturer. See DNB.

p.45, line 14   Erasmus describes the speaking machine and his theory of phonetics in the Additional Notes to *The Temple of Nature*, pp.107–120, as indicated in C.D.'s note 28. The jokey agreement with Matthew Boulton is stuck into the Commonplace Book, p.210.

p.46, line 10   For more details of the 'artesian well' and its principles, see 1999 *Life*, pp.194–6.

p.46, line 12   The iron plate is now in the Derby Museum. The inscription has a fifth line , 'Philos. Trans. V.75', referring to Erasmus's paper in the *Philosophical Transactions*, Vol.75, pp.1–7, 1785. But this line is irrelevant to the message.

The four main lines may loosely be translated as: 'Erasmus Darwin, in the year 1783, extracted water from a small borehole. It flows and shall flow'. All well and good perhaps, though *labitur* means 'it sinks, slides or glides' rather than 'it flows'. But why did Erasmus choose to record this message, and no other, in tablets of stone (or, rather, letters of iron)? When young, Erasmus was expert at word-play, in English and Latin, and produced a flow of 'enigmas' (see DAR 267). Perhaps he is teasing us here with Latin puns?

The first word *Terebella* is provocative: arguably, it should be *Terebella*, from the diminutive of *Terebra*, a gimlet or auger. So 'Erasmus Darwin . . . extracted water with a small gimlet' is an alternative and more amusing translation. His actual borehole was 2½ inches in diameter. The inscription as given by C.D. has *Terrebello* as the first word, and this misprint points to further puns. *Terra bella*, 'from the good earth', is better than *Terrebello* in both Latin and English. Also possible is *Terra belle*: 'extracted water well from the earth', where 'well' is short for 'elegantly', as well as 'a well'.

The last line also suggests many puns, of which the sharpest is *Labitor et libator*, 'let it be sunk and let it be drunk', a neat twist in Latin, a perfect rhyme in English, and a stronger punch line (though 'be sunk' should really be 'sink'). Erasmus would have seen all these puns in a flash. If I am wrong, I apologize for wasting space: but don't forget that there is no way of correcting misprints on an iron plate.

p.46, line 29   For the narrow canal project, see 1999 *Life*, pp.101–2.

p.47, line 4   For Erasmus's many bright ideas in meteorology, including his discovery of how clouds usually form, see 1999 *Life*, pp.148–9, 216–17, 226–8, 269–71.

p.47, line 7   Such wind-vanes were fitted to several houses by the clockmaker and inventor John Whitehurst (1713–1788), who was a great friend of Erasmus for thirty years and encouraged his inventiveness. Whitehurst probably installed Erasmus's wind-vane. See M. Craven, *John Whitehurst of Derby* (Mayfield Books, 1996), pp.28–30.

p.47, line 14   In his role as author, C.D. seems to condemn a letter as 'uninteresting' unless it is 'serious', scientifically, medically or socially. The first four here are all about deaths, and C.D. does not choose any of the brilliant and bantering letters of Erasmus that have proved most popular in recent years. All the letters quoted are now available in manuscript. The transcriptions are excellent, apart from the paragraphing, and I have left the letters unchanged except for inserting new paragraphs and correcting a few misprints.

p.47, line 16   Rousseau stayed at Wootton Hall in the Weaver Hills of north-east Staffordshire for nearly a year. As C.D. says, there is no known surviving correspondence between Erasmus and Rousseau.

p.48, line 1   A more complete copy of the letter about the dead baby is in *Letters*, pp.41–3. The recipient is not known, but the letter as sent (with the final page missing) is in the possession of the Meynell family. It differs little from the draft version used by C.D.

p.48, line 25   Note that Erasmus always uses the spelling 'Wedgewood'.

p.49, line 1   The great canal engineer James Brindley died of diabetes on 27 September 1772, aged 56. Josiah Wedgwood and Erasmus both knew him well. Erasmus's 'eulogium' was eventually published in *The Botanic Garden*, Part I, Canto III, lines 329–36. See *Letters*, pp.65–6 and 1999 *Life*, pp.111–12. Mr Stanier was Brindley's doctor. Mr 'Henshaw' is Hugh Henshall (1733/4–1816), Brindley's brother-in-law, assistant and successor. In the last paragraph of the letter, the words 'I wish you would . . . me' are printed in italics in the proofs, and are underlined in the manuscript, DAR 227.1:39.

p.49, line 17   Wedgwood's close friend and business partner Thomas Bentley (1730–1780) died suddenly on 26 November 1780. It was a devastating blow for Wedgwood, who had relied entirely on Bentley to handle the London business of the firm of Wedgwood and Bentley.

By a strange coincidence, Erasmus's own life was thrown into turmoil simultaneously, because Colonel Pole died on 27 November and Erasmus had to face the probability that Elizabeth would be snatched from him by some handsome, dashing and wealthy rival. (C.D. was of course unaware of this coincidence.) For further notes, see *Letters*, pp.103–4. The manuscript is now available as DAR 227.1:61. Note that Erasmus uses the spelling 'Bently'.

p.50, line 20 'Your late loss' refers to the death of Edgeworth's daughter Honora (1774–1790), who had died of consumption at the age of 15. For the complete letter with notes, see *Letters*, pp.201–3, for the manuscript, see DAR 218. The text given by C.D. is correctly transcribed from Edgeworth's *Memoirs*, but the editor of the *Memoirs*, Maria Edgeworth, changed the word 'Forgetfulness' in the manuscript to 'Time'. Probably she felt uneasy at the idea of forgetting her dead half-sister.

p.51, line 1 In this letter Erasmus is tactfully trying to persuade Josiah to abandon his erroneous 'theory of freezing steam', and offers to submit the matter to the world's leading authority, Dr Joseph Black (1728–1799), famous for discovering and defining latent heat and specific heat. 'Mr Robert' is Erasmus's son Robert, who was now at the Edinburgh Medical School, where Black was one of his teachers. This is the longest of Erasmus's letters, and includes several speculative ideas, as Charles remarks. The complete letter is available in manuscript as DAR 227.1:75, and has been printed, with other related letters, in *The Correspondence of Charles Darwin*, Vol.9, pp.407–15 (1994).

p.51, line 25 The half-sentence 'Therefore I suppose . . . distances' is underlined in the manuscript.

p.53, line 4   Erasmus foreshadows his law of adiabatic expansion (1788) when he states that 'air when it is mechanically expanded always attracts heat from the bodies in its vicinity'.

p.53, line 26   The word 'not' appears in the manuscript (DAR 227.1:113), but is probably a mistake.

p.54, line 2   Dr William Small (1734–1775) was a man of immense knowledge and kindness, who arrived in Birmingham in 1765, soon became the best friend of Matthew Boulton and Erasmus, and acted as chief nurturer of the Lunar group until his early death. For Small, see G. Hull, *Journ. Roy. Soc. Medicine*, Vol.90, pp.102–5 (1997).

p.54, line 16   The 'young man' is identified in the manuscript (DAR 227.1:132) as 'Salt's son'. He is probably Dr John Butt Salt (1768–1804), son of the Lichfield surgeon Thomas Salt and elder brother of Henry Salt (see DNB). The 'young man' did take a degree at Edinburgh, and so Erasmus's letter was unnecessary: but it shows how much trouble he would take to help Robert. In 1795 Erasmus recommended Salt to Boulton and in 1796 to Watt: see *Letters*, pp.278 and 292.

p.54, line 21   'Mr Green of Lichfield' is Richard Greene (1716–1793), the apothecary who created a museum much admired by Dr Johnson and others. See DNB.

p.55, line 27   'Mr Howard' is Robert's uncle Charles Howard (1742–1791), the younger brother of Polly.

p.56, line 25   'Dr Crawford' is Adair Crawford (1748–1795). See DNB and DSB (Vol.15). For further notes on this letter, see *Letters*, p.221.

p.56, line 27   James Hutton (1726–1797) was a close friend-at-a-distance of Erasmus. In 1774 Hutton stayed at Erasmus's house, and they met again in Edinburgh in 1778, when Hutton supervised the monument for Erasmus's son Charles. Their letters seem to have been lively, with Lunar-type banter, but only one on each side has survived. The date of this letter is probably about 1782. For Hutton's work, see D. R. Dean, *James Hutton and the History of Geology* (1992).

p.57, line 12   Erasmus was, as Charles suggests, in favour of the American Revolution, in which his old friend Benjamin Franklin (1706–1790) was so deeply involved. The quotation is from the manuscript DAR 227.1:70.

p.57, line 19   For Anna's correspondent Dr Thomas Whalley (1746–1828), a wealthy literary man, see DNB.

p.57, line 28   For Lady Charleville (1762–1851), see *The Collected Letters of Joanna Baillie*, Vol.2, ed. J. B. Slagle (1999), p.1040.

p.58, line 8   According to Anna Seward (*Memoirs of Dr Darwin*, p.76): 'Where Dr Johnson was, Dr Darwin had no chance of being heard, though at least his equal in genius, his superior in science'.

p.58, line 10   'Mr Seward' is Anna's father, Revd Thomas Seward (1708–1790), canon residentiary of Lichfield Cathedral, who wrote poems published in Dodsley's *Miscellany*, as well as editing Beaumont and Fletcher's plays. For Seward, see F. Swinnerton, *A Galaxy of Fathers* (1966), pp.35–47. The text of the poem is taken from Sir Francis Darwin's collection of his father's poems (UCL Pearson Papers 577, p.48).

p.58, line 19   For this quotation see DAR 227.5.12.

p.59, line 12   This quotation is apparently not from Martial.

p.59, line 24   Dr Caleb Hardinge (1701–1775) was notorious for speaking his mind, and for witty but wounding remarks. See J. Nichols, *Illustrations of the Literary History of the 18th Century*, Vol.3, pp.1–224, for the Hardinge family, and p.4 for Caleb. Erasmus's reply was sharper than C.D. realised, because Dr Hardinge himself stuttered (see J. Nichols, *Literary Anecdotes . . .*, Vol.8, p.523).

p.59, line 28   Charles's summary of Erasmus's intellectual powers is generous and perceptive. He had 'vividness of imagination', 'great originality of thought', 'the true spirit of a philosopher' and 'uncommon powers of observation', all applied in a 'surprising' diversity of subjects. Reading his own words, Charles may have thought, 'Well, I suppose he was probably the most talented man of his time. But you won't catch me saying that, and being accused of family boastfulness'. And so he continued to treat Erasmus as a not very important person, whose life was not of much intrinsic interest.

p.60, line 10   Keir's letter is available as DAR 227.6:76.

p.60, line 18   'Mr Day' is Thomas Day (1748–1789), the richest, most virtuous and most eccentric member of the Lunar group, and a good friend of Erasmus, Edgeworth and Keir. Day failed in his experiment à la Rousseau of educating an orphan girl to become his wife. After finding a wife elsewhere, he retired from the world, running a farm on barren ground and expending his fortune in paying his labourers. He was author of a famous anti-slavery poem in 1773 and of the children's book *Sandford and Merton* (1783–8), said to have been the most-read modern book of the nineteenth century: at least 131 editions are recorded. The best biography of Day is P. Rowland, *The Life and Times of Thomas Day* (1996).

p.61, line 4   For Erasmus's letter, see DAR 227.1:146.

p.61, line 20   The grave of Doctor can still be seen at Breadsall Priory, which is now a hotel.

p.62, line 15   The first stanza given by C.D. is as written out by Sir Francis Darwin's wife Jane in the Commonplace Book, pp.171–4. However, she gives the first line of the last stanza as: 'What parent power all great and good'.

There is another version in the UCL Pearson Papers 577, pp.28–9.

p.62, line 25   Dr Andrew Kippis (1725–1795) was a nonconformist pastor, biographer and teacher at Hackney College. See DNB. With three others, he compiled a widely-used *Collection of Hymns . . .* (1795). Erasmus's hymn is on pp.261–2, and is entitled 'Prosperity and Adversity'. C. D. probably based his text on that written out by Sir Francis Darwin in the Commonplace Book, pp.165–6.

p.62, line 26   Dr James Martineau (1805–1900), brother of Harriet Martineau and a friend of the Darwins, edited two volumes of hymns. See DNB.

p.63, line 12   Note that C.D. (incorrectly) calls all the children of Erasmus's second marriage 'stepsons' or 'step-daughters'. Here the 'step-daughter' is Violetta.

p.63, line 19   These two sentences are interesting because Charles himself suffered similar apocryphal stories.

p.63, line 28   C.D.'s impartiality seems unassailable: first the blemishes, then the virtues. No family favouritism here.

p.64, line 12   The 'stepson' here is presumably Francis: the other two 'stepsons' and the real stepson all died before 1830.

p.64, line 19   Mary Parker (1753–1820) was Erasmus's mistress. Their two illegitimate daughters, Susanna Parker (1772–1856) and Mary Parker (1774–1859), jointly conducted the boarding school for girls at Ashbourne from 1794 until Susanna's marriage to Henry Hadley in 1809. Thereafter Mary Parker was in charge of the school until her retirement in 1827. Both seem to have been 'admirable ladies', as C.D. says. One of her Hadley nieces said of Susanna: 'What I owe to her no tongue can tell'. Mary was a respected and wealthy Ashbourne resident, who outdid her father by appearing among the 'Nobility, Gentry and Clergy' in the 1821 Directory of Derbyshire. For more information see 1999 *Life*, pp.281–4, 357–8 and 369.

p.64, line 28　Lord Thurlow (1731–1806) was a powerful legal and political figure, being Lord Chancellor from 1778 to 1792, apart from a brief gap in 1783–4. See pp.473–678 in J. Campbell, *Lives of the Lord Chancellors*, Vol.5 (Murray, 1846), as cited by C.D.

p.65, line 2　The case is indeed 'a very odd one', and C.D.'s analysis is persuasive and well informed.

p.65, line 30　'We will now turn to the favourable side of his character', says C.D. Yet he then proceeds to quote all the calumnies against Erasmus, and, although he refutes them vigorously, the overall effect is rather negative. Most readers would welcome Henrietta's considerable cuts in the calumnies.

p.66, line 28　The three letters written in 1792–3 are available in manuscript as DAR 227.1:143, DAR 227.1:146, and DAR 227.6:36.

p.68, line 8　Here *secundum artem* (according to art) means 'with proper legal wording'.

p.68, line 27　Keir's *invitâ Minervâ* implies 'at someone else's request'. For Keir's letter, see DAR 227.6:76 (again).

p.69, line 2　'Mr Smith' is John Raphael Smith (1752–1812), son of Thomas Smith of Derby. See Bryan's *Dictionary of Painters and Engravers* (1905) v 94. The quotations are from Erasmus's letter to Robert of 6 May 1797, DAR 227.1:159. Smith engraved Wright's 1792 portrait, which appears in the 1879 book. But this 2002 book has Wright's 1770 portrait as frontispiece.

p.69, line 10　Edgeworth's comment on the unsmiling portrait is echoed by others. Erasmus's friend W. B. Stevens (1756–1800) called Wright's 1792–3 portrait 'a strong but severe likeness. His countenance is seldom without a Smile playing round it'. (*Journal*, 1965, p.65). Even the ever-critical Anna Seward says 'his sunny smile . . . rendered . . . that exterior agreeable, to which beauty and symmetry had not been propidous'. (*Memoirs of the Life of Dr Darwin*, 1804, p.2.)

p.69, line 18   Lady Charlotte Finch (1725–1813) was governess to all the children of George III and Queen Charlotte from 1762, for nearly 40 years. Her funeral procession included five carriages of Royal Dukes. According to Baroness Bunsen (*Memoirs*, p.68), it was Augusta Fielding who was treated by Erasmus. For pictures of Lady Charlotte and Augusta, see P. Finch, *History of Burley-on-the-Hill* (1901) i 20 and i 311.

p.69, line 33   Dr Fox is Dr Francis Fox (1759–1833) of Derby, father of the engineer Sir Charles Fox (1810–1874), who was responsible for the building of the Crystal Palace, Waterloo Station, etc. Mr Hadley was Henry Hadley (1762/3–1830), who married Susanna Parker in 1809. Fox and Hadley conducted the post-mortem on Erasmus (DAR 227.8:15), and published their denial in the *Derby Mercury*.

p.70, lines 1–13   C.D.'s original draft of these lines is available as DAR 212:6.

p.70, line 11   Robert's comment can be found in DAR 227.5:12.

p.70, line 21   Anna Seward's book, called *Memoirs of the Life of Dr Darwin*, is not a connected narrative: 'Comments on the Life and Work of Dr Darwin, with digressions' would be a more accurate title. She says virtually nothing about his 21 years at Radburn and Derby: instead she offers 216 pages of commentary on *The Botanic Garden* – the best critique yet of the poem.

p.71, line 22   The manuscript of Erasmus's letter of 30 December 1799 is available as DAR 227.1:165.

p.71, line 32   The complete letter from Emma Darwin (1784–1818) to Mrs Susannah Darwin (1765–1817), Charles's mother, dated 13 February 1804, is available as DAR 227.7:42.

p.72, line 21   The letter of Robert to Anna Seward of 10 February 1804 is available as DAR 227.4:9; and Erasmus's letters to Robert of 8 February 1800 and 26 February 1800 as DAR 227.1:166 and DAR 227.1:167.

p.73, line 1   Anna's apology was published alongside the review of her book in the Edinburgh Review, April 1804, pp.236–7.

p.73, line 21   Erasmus's letter of 8 June 1798 is available as DAR 227.1:161.

p.74, line 6   I have suggested (1999 Life, pp.326–7) that the suicide may possibly have been an accident.

p.74, line 16   For Erasmus's letter of 28 November 1799 (or, more probably, 23 November), see DAR 227.1:164.

p.74, line 31   For the full text of Anna Seward's letter, see DAR 227.6:73.

p.76, line 1   C.D.'s explanation of Anna's hostility is open to some objections. Anna was only fourteen in the months before Erasmus's first marriage, so Henrietta's deletion of the sentence is justified. The idea that Anna had designs on Erasmus after Polly's death in 1770 seems more likely. But C.D. did not know that in 1767 Anna began her lifelong association with John Saville (1736–1803), and this friendship, though never carnal, was passionately felt. When she wrote the book, in 1803 after Saville's death, she was depressed, and bitter about her life; she might have been taking revenge, as C.D. suggests.

p.76, line 20   Here comes another of Charles's little jokes. The calumnies published by Mrs SchimmelPenninck, he says, 'are hardly worth notice': whereupon he goes through them quite fully (though much of the material

was rightly deleted by Henrietta). Mrs
SchimmelPenninck was unreliable, as C.D. says, and
also mendacious: 'she broke off eleven marriages',
according to her nephew Sir Francis Galton (it was
thirteen according to another relative). Charles's
comments on the calumnies are well judged.

p.78, line 17   I have not located this 'Guide-Book to Derbyshire'.

p.78, line 20   Keir's letter to Robert is available as DAR 227.6:81.

p.78, line 29   'The old divine, Hooker' is the theologian Richard
Hooker (1554?–1600), author of *Laws of Ecclesiasticall
Politie*.

p.79, line 6   The complete original letter from the 'judicious' James
Keir is available as DAR 227.6:76.

p.79, line 23   Sir Brooke Boothby (1744–1824), poet and author,
was a close friend of Erasmus, and collaborated with
him in the translations from Linnaeus in the 1780s. For
Boothby, see DNB and a forthcoming biography, *Sir Brooke
Boothby* by J. Zonneveld, who mentions that Boothby
met Charles – when Charles was four months old!

p.79, line 31   The 'old woman living in Birmingham' is specified in
the original letter (DAR 227.1:146) as 'Mrs Day, No 21
Prospect Row, Coleshill-Street, Birmingham' – the
39-year-old former Mary Parker, who had not been
forgotten by Erasmus. In 1782 she had married Joseph
Day, variously described in Birmingham Directories as
'Watch-key and sealmaker' (1774–7), 'Toy-maker'
(1777–83) and 'Stamper' (1787).

p.80, line 4   'Mrs Darwin' here is Erasmus's first wife Polly. Her
brother, Charles Howard (1742–1791) was later subject
to more frequent intoxication. See *Letters*, p.218.

p.81, line 7   This 'Mr Day' is of course Thomas Day (1748–1789):
see p.124. Erasmus's letter about him is available as
DAR 227.1:119.

p.81, line 14 'The love of woman . . .' This sentence is a masterpiece of delicate phrasing, but not delicate enough to escape Henrietta's axe. Many of Erasmus's love-poems are quoted in part in the 1999 *Life* (pp.125, 127, 139–40, 156–7, 168, 171); three are quoted in full in the *Letters* (pp.77, 86–7, and 91–3); and the complete set is available at UCL Library, Pearson Papers 577.

p.81, line 28 Boothby's verses are from his *Sorrows Sacred to the Memory of Penelope* (1796), p.71.

p.82, line 3 Lord George Cavendish (1728–1794), his nephew the fifth Duke of Devonshire (1748–1811), and the Duke's wife Georgiana (1757–1806), all consulted Erasmus about medical problems. See *Letters*, pp.130-3, 263-5, 325–9.

p.83, line 3 The stories came to Charles through Mrs Elizabeth Anne Wheler (1808–1906), the daughter of Violetta Galton. See DAR 210.14:23.

p.83, line 21 It seems that the dispensary did serve as the precursor of the Derby Infirmary, designed 20 years later by Erasmus's friend William Strutt (1756–1830) and remarkable for being fireproof and well-ventilated.

p.83, line 23 Charles's diatribe against the opponents of vaccination (inevitably deleted by Henrietta) is a rare breaking out of his feelings, perhaps with a tacit side-swipe at the opponents of evolution?

p.83, line 31 An earlier Philosophical Society in Derby had faded away. The new Society was begun by Erasmus in March 1783 (*Letters*, p.128), when Susannah Wedgwood (Charles's mother) went to one of the meetings. See 1999 *Life*, pp.188–9.

   Later in the year Erasmus constructed a hydrogen balloon 5 ft in diameter. It was launched as an 'event' for the Society on 26 December and flew for 30 miles.

Erasmus thus became the first Englishman to fly such a balloon.

The Derby Philosophical Society continued for 70 years. William Strutt was the second President. See 1999 Life, pp.196 9.

p.84, line 2   Erasmus's presidential address is not in the proofs. Charles was sent a copy of it by Reginald Darwin on 4 August 1879 (DAR 99:160-1), and immediately decided to include an excerpt. (I have reduced the number of words in italics, but the emphasis is preserved.)

p.84, line 28   The 'positive assertions' come from Anna Seward, who said 'it is well known that Dr Darwin's influence and example have sobered the county of Derby' (*Memoirs of the Life of Dr Darwin*, p.5); and from Maria Edgeworth, who said he 'persuaded most of the gentry in his own and the neighbouring counties to become water drinkers' (Edgeworth, *Memoirs* ii 82).

All this is difficult to believe, but the fact that it *was* believed signals Erasmus's great prestige as a doctor. Certainly, he was severe in telling Derbyshire's leading aristocrat (the Duke of Devonshire) to cut his drinking. See *Letters*, pp.130–3.

p.85, line 7   Here I have inserted 'Much earlier' and deleted 'also'.

p.85, line 13   'Dr Richardson', later Sir Benjamin Ward Richardson (1828-1896) was an eminent physician and sanitary reformer, and also a poet. See DNB and DSB. He had earlier planned to write a life of Erasmus.

p.85, line 21   'Mr Sneyd' is John Sneyd of Belmont (1734–1809). Anna Seward has a gushing tribute to Sneyd in *Memoirs of Dr Darwin*, pp.100–4.

p.86, line 1   Mr Sneyd's letter to Mrs Elizabeth Darwin, dated 22 February 1804, is available in manuscript as DAR 227.7:44. The 'stepson' mentioned was Francis Darwin, the son of Erasmus and Elizabeth.

p.86, line 13   Erasmus was much impressed by the selfless labours of John Howard (1726–1790), and so was Charles. The verses about Howard are not in the proofs and were added later. The lines quoted from Canto II are 439–58, 463–6 and 469–72. (The omitted lines 467–8 are a digression, and I have not marked the gap.)

p.87, line 15   C.D. indicated on the proof that he wished to quote the passage in *Phytologia*, so I have inserted it within square brackets.

p.87, line 20   Erasmus was ahead of his time: but so was Charles; few Victorian authors saw slavery as 'this great national crime'.

p.87, line 22   *Colonel Jack* (1722) is one of Defoe's lesser-known picaresque adventure stories.

p.88, line 4   C.D. hits the target with the slave story from Lecky's *History*.

p.88, line 6   The 'notes made by my father' (DAR 227.5:12) are the source of the personal comments in this short section. (I considered splitting it up and fitting in the fragments elsewhere: but the results were not good, and I left it untouched.)

p.88, line 17   Erasmus broke the patella of his knee in his carriage accident in July 1768, but I have no evidence that he broke it twice. Hence my query.

p.88, line 21   Erasmus was pitted with smallpox when young, according to Anna Seward and other witnesses. But his daughter Violetta and his step-daughter Millicent Gisborne both told their daughters that Erasmus's 'complexion' was 'clear and good in his later life', or even 'beautiful, like a child's'. Elizabeth Wheler wrote to tell Charles of this on 10 January 1880, DAR 99.197–8.

p.88, line 25   Charles undervalued Erasmus initially; but, as he proceeded, he became more and more impressed, as this final tribute shows. Unfortunately Henrietta decided to delete it all. Why? She was obviously incensed by the link between the calumnies and 'the state of Christian feeling' in the early 1800s. Charles's 'hope that nothing of the kind now prevails' was immediately shattered by her action. It would have been enough for her to have deleted the two last sentences. Why delete all the rest? I presume she may have thought Charles's tribute was too favourable an envoi for a man who had two illegitimate children and did not believe in revelation.

p.89, line 16   Erasmus, Elizabeth and their family moved out of Derby to Breadsall Priory about 25 March 1802. Elizabeth's foster-mother Mrs Susan Mainwaring was staying with them and walking in the garden on 17 April.

p.89, line 21   The 'three weeks' here is my correction of the 'two years' in the 1879 book. (The phrase beginning 'where he . . . ' does not appear in the proofs, and was probably an addition made by Henrietta.)

p.89, line 23   Erasmus's letter to Edgeworth is printed in full, with comments, in *Letters*, pp.338–9.

p.89, line 29   This account of Erasmus's death is based on the letter from Robert Darwin to Edgeworth of 1 May 1802, DAR 218:C1.

p.91, line 3   The quotation is from Erasmus's letter to Edgeworth of 15 March 1795: *Letters*, p.279.

p.91, line 6   The cause of death was probably a lung infection. See G.C. Cook and D. King-Hele, *Notes and Records of the Royal Society*, Vol.52, pp.261–5 (1998). The inconclusive post-mortem (DAR 227.8:15) is printed in this paper.

p.91, line 8   The date of Charles's visit to Breadsall Priory is not known. Perhaps it was on one of his journeys between

Shrewsbury and Cambridge in 1828–31? Elizabeth died on 5 February 1832.

p.91, line 11   The actual memorial inscription has 'M.B,' rather than 'M.D.' However, Erasmus styled himself 'M.D., F.R.S.' in both *Zoonomia* and *Phytologia*, so I assume that the 'M.B.' was an inscription misprint. C.D. probably assumed the same: the proofs, the 1879 book and the second edition (1887) all have 'M.D.'

p.91, line 13   The date of Erasmus's death was misprinted as '10th April' in the proofs, the 1879 book and the second edition. (C.D. was aware of this error; letter to Krause, 13 February 1880.)

p.92, Note 1   The reference given wrongly by C.D. is an article by W. C. Wells (1757–1817) disputing Erasmus's ideas on giddiness in *Zoonomia*. (Ironically, Wells was another early advocate of natural selection: see C. D. Darlington, *Darwin's Place in History* (1959), pp.85–90.) Charles was much helped in tracing references by his son George, who wrote on one sheet (DAR 227.8:65) summaries of both Wells's article and the correct one by Duncan, dated 1801: C.D. made the mistake of giving a reference to the first when intending to refer to the second.

p.92, Note 5   'Dr Sieveking' is Sir Edward Sieveking (1816–1904), a leader in English medicine, whose father was prominent in Hamburg, where Reimarus lived. See DNB.

p.92, Note 10   The second edition of Edgeworth's *Memoirs*, cited by C.D., is very rare in Britain. So I have added page references to the first edition, of which there is a modern reprint.

Appendix A
Chronology of Erasmus Darwin's Life

1723/4 Jan 1	Father, Robert Darwin, marries Elizabeth Hill. They live at Elston Hall, near Newark, Nottinghamshire.
1731 Dec 12	Born at Elston Hall. Seventh child, fourth son.
1731/2 Jan 13	Baptised at Elston Church.
1732–41	Living at Elston Hall with family. Going to school at Elston. Makes experiments in mechanics and electricity.
1741–50	At Chesterfield School: classical and literary education.
1747–50	Writes many poems and letters in verse to brothers and sister Susannah.
1750 June 30	Admitted to St John's College, Cambridge, with brother John. Awarded Exeter Scholarship.
1750 Oct –1753 June	At St John's College, with some vacations at home. Writes many poems and letters in verse.
1751 May	First published poem, 'Death of Prince Frederick'.
July	Second published poem, 'To Mr Gurney'.
? 1753 Jan–Apr	Attends W. Hunter's lectures on anatomy in London.
1753 Oct	Travels to Edinburgh Medical School.
1754	At Edinburgh. Meets James Keir.
1754 Sept	Father, Robert Darwin, dies aged 72.
? 1755 to June	At Cambridge. Takes M.B. degree. Meets John Michell.
? 1755 Oct –1756 June	At Edinburgh Medical School. Probably qualifies as M.D. Meets Albert Reimarus.

1756 July	At Elston Hall.
1756 Aug–Oct	At Nottingham. Medical practice fails.
1756 Nov 12	Arrives at Lichfield; starts successful medical practice.
1757	Lives in lodgings in Cathedral Close; then at a house in Lichfield. Sister Susannah acts as housekeeper. Meets Anna Seward (then 14).
1757 May	Paper on 'Ascent of Vapour' published in *Philosophical Transactions* of Royal Society.
1757 Dec 30	Marries Mary (Polly) Howard.
1758	Acquires house at edge of Cathedral Close, Lichfield: enlarges and improves it. Now open to visitors as 'Erasmus Darwin's House'.
1758 Sept	First son Charles born.
1759 Oct	Second son Erasmus born.
1759	Devises improved technique for carriage steering.
1757–60	Meets Matthew Boulton, John Whitehurst, Benjamin Franklin. Probably meets Samuel Garbett, Robert Bage, James Brindley.
1761 April	Elected Fellow of the Royal Society.
1762 Oct	Dissects 'malefactor'.
1763 Nov	Daughter Elizabeth born: dies March 1764.
? 1763	Sketches design for steam car. Suggests joint project with Boulton to construct it.
1764	Entrepreneur of Wychnor Ironworks, with Garbett, Bage, John Barker.
1765	Meets Josiah Wedgwood: involved with him in promotion of Grand Trunk Canal. Writes canal pamphlet.
1765 May	William Small settles in Birmingham. Lunar Society begins to develop.
1766 May	Third son Robert born: he later becomes father of Charles Darwin (1809–1882).
1766 Summer	Meets Richard Edgeworth, J. J. Rousseau.
1766 Nov	First sign that his wife Polly is suffering illness.

1767 July	Son William Alvey born: dies in August.
1767	Meets James Watt, Joseph Wright, Samuel Johnson.
1767 Nov	Student friend James Keir comes to live in the area.
1768 Mar	Meets Thomas Day.
1768 July 11	Carriage accident: breaks patella of knee.
1769	Illness of his wife Polly worsens.
1770 June 30	Polly dies, of gallstones and liver disease.
1770 July	Sister Susannah returns as housekeeper.
1770	Portrait painted by Joseph Wright. Begins writing *Zoonomia.*
1771	Adds evolutionary motto *E conchis omnia* to bookplate.
1771	Liaison with Mary Parker begins.
1772 May 16	Susanna Parker born.
1772	Constructs speaking machine.
1774 May 20	Mary Parker born.
1774 June	Meets James Hutton. James Watt settles in Birmingham.
1774	Forms Lichfield literary circle, with Anna Seward, Brooke Boothby.
1775 Feb 25	Dr Small dies, aged 40, probably of malaria.
1775	Falls in love with Elizabeth Pole.
1776	Begins creating botanic garden at Abnalls, Lichfield.
1776	Begins writing Commonplace Book.
1776	Eldest son Charles enters Edinburgh Medical School.
1776–79	Writes poems to Elizabeth Pole.
1777	Sketches inventions of canal lift and artificial bird.
1778 May 15	Eldest son Charles dies, aged 19, of infected wound.
1779	Constructs copying machine; and horizontal windmill. Sketches multi-mirror telescope.
1779	Completes Lichfield botanic garden. Forms Lichfield Botanical Society.
1780 Sept	Joseph Priestley settles in Birmingham.
1780 Nov 27	Colonel Pole dies, aged 63.
1781 Mar 6	Marries Elizabeth Pole.

1781 Mar	Leaves Lichfield and goes to live at Radburn Hall, near Derby.
1781 April	Goes to London with Elizabeth; they stay six weeks. Meets Sir Joseph Banks.
1782 Jan 31	Son Edward born. Son Erasmus becomes solicitor in Derby.
1783 Mar	Forms Derby Philosophical Society.
1783 Apr 23	Daughter Violetta born.
1783	Works on Linnaean translation.
1783 Oct	Moves from Radburn Hall to house in Full Street, Derby. Drills 'artesian' well there.
1783 Dec 26	Flies 5-ft hydrogen balloon, which travels 30 miles.
1784 Aug 24	Daughter Emma born.
1785	Linnaean translation, *System of Vegetables*, published.
1785 Feb	Travels to London for Arkwright patent trial. Travels there again in June, with wife Elizabeth.
1785	Paper on 'An artificial spring of water' published in *Philosophical Transactions* of the Royal Society.
1786 June 17	Son Francis born. Son Robert starts as doctor in Shrewsbury (Aug).
1787 Sept 5	Son John born. Linnaean translation, *Families of Plants*, published.
1788 Feb 18	John Whitehurst dies, aged 74.
1788	Paper on 'Mechanical expansion of air' published in *Philosophical Transactions* of the Royal Society.
1789 Apr	Publication of *The Loves of the Plants* (Part II of *The Botanic Garden*).
1789 Apr 10	Son Henry born: dies 1790.
1789 Apr 29	Sister Susannah dies, aged 60.
1789 Sept 28	Thomas Day dies in accident, aged 41.
1789 Oct	Josiah Wedgwood sends first good copy of Portland Vase.
1790 July 5	Daughter Harriot born. Benjamin Franklin dies, aged 83 (Apr).

1791 July 14	Birmingham riots: Priestley's house burnt. Lunar Society declines.
1791 Dec	Derby Society for Political Information formed.
1792 June	Publication of *The Economy of Vegetation* (though dated 1791). Part I of *The Botanic Garden*. E. D. now recognized as the leading English poet.
1793	Second portrait by Joseph Wright completed.
1793 Dec	Seditious libel trial over Derby Society's 'Address'.
1794 May	Publication of *Zoonomia*, Volume I. Chapter 39 openly advocates biological evolution.
1794	Susanna and Mary Parker start girls' boarding school at Ashbourne.
1795 Jan 3	Josiah Wedgwood dies, aged 64.
1796 Jan 24	Has long talk with S. T. Coleridge at Derby.
1796	Publication of *Zoonomia* Volume II. E. D. now regarded as leading medical writer.
1797	Mother, Elizabeth, dies, aged 95. Joseph Wright dies, aged 62.
1797	Publication of *Female Education in Boarding Schools*.
1798 Apr	Satirized by the Anti-Jacobin with poem, 'Loves of the Triangles'.
1798	Criticized by Wordsworth and Coleridge in *Lyrical Ballads*.
1799 Dec 29	Son Erasmus drowned in River Derwent, aged 40. Suicide or accident?
1800 Apr	Publication of *Phytologia: or the Philosophy of Agriculture and Gardening*.
1801 Mar	Ill with pleurisy or pneumonia. Third edition of *Zoonomia* published.
1802 Mar 25	Family moves to Breadsall Priory, 4 miles north of Derby.
1802 Apr 18	Dies at Breadsall Priory, aged 70, probably of lung infection.
1802 Apr 24	Buried at Breadsall Church.

1803	Publication of *The Temple of Nature; or, The Origin of Society.*
1804 Feb	Publication of Anna Seward's *Memoirs of the Life of Dr Darwin.*
1809 Feb 12	Grandson Charles born.
1832 Feb 5	Widow Elizabeth dies, aged 84.
1848 Nov 13	Son Robert dies, aged 82.
1874 Feb 12	Daughter Violetta dies, aged 90.

Appendix B
Selective Family Tree

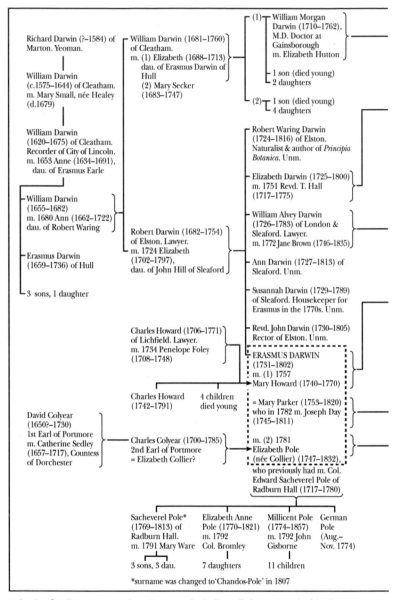

Selective family tree, centred on Erasmus, including all the names cited in the text.

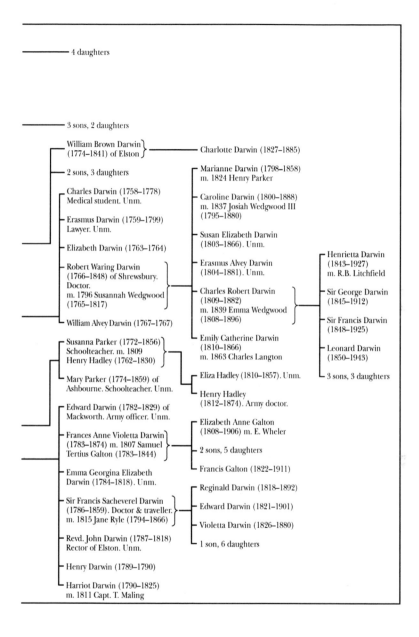

4 daughters

3 sons, 2 daughters

William Brown Darwin (1774–1841) of Elston ——— Charlotte Darwin (1827–1885)

2 sons, 3 daughters

Marianne Darwin (1798–1858) m. 1824 Henry Parker

Charles Darwin (1758–1778) Medical student. Unm.

Caroline Darwin (1800–1888) m. 1837 Josiah Wedgwood III (1795–1880)

Erasmus Darwin (1759–1799) Lawyer. Unm.

Susan Elizabeth Darwin (1803–1866). Unm.

Elizabeth Darwin (1763–1764)

Erasmus Alvey Darwin (1804–1881). Unm.

Robert Waring Darwin (1766–1848) of Shrewsbury. Doctor. m. 1796 Susannah Wedgwood (1765–1817)

Charles Robert Darwin (1809–1882) m. 1839 Emma Wedgwood (1808–1896)

Henrietta Darwin (1843–1927) m. R.B. Litchfield

Sir George Darwin (1845–1912)

William Alvey Darwin (1767–1767)

Sir Francis Darwin (1848–1925)

Emily Catherine Darwin (1810–1866) m. 1863 Charles Langton

Leonard Darwin (1850–1943)

Susanna Parker (1772–1856) Schoolteacher. m. 1809 Henry Hadley (1762–1830)

Eliza Hadley (1810–1857). Unm.

3 sons, 3 daughters

Mary Parker (1774–1859) of Ashbourne. Schoolteacher. Unm.

Henry Hadley (1812–1874). Army doctor.

Edward Darwin (1782–1829) of Mackworth. Army officer. Unm.

Elizabeth Anne Galton (1808–1906) m. E. Wheler

Frances Anne Violetta Darwin (1783–1874) m. 1807 Samuel Tertius Galton (1783–1844)

2 sons, 5 daughters

Emma Georgina Elizabeth Darwin (1784–1818). Unm.

Francis Galton (1822–1911)

Sir Francis Sacheverel Darwin (1786–1859). Doctor & traveller. m. 1815 Jane Ryle (1794–1866)

Reginald Darwin (1818–1892)

Edward Darwin (1821–1901)

Revd. John Darwin (1787–1818) Rector of Elston. Unm.

Violetta Darwin (1826–1880)

Henry Darwin (1789–1790)

1 son, 6 daughters

Harriot Darwin (1790–1825) m. 1811 Capt. T. Maling

Appendix C
Selected Books and Papers

This list records the first editions of Erasmus Darwin's own books. (Later editions and translations, and his papers, are given in 1999 *Life*, pp.401–3.) I have added a selection of biographical books about Erasmus Darwin and his friends, and three books about Erasmus's literary work and influence. Finally, there is a very sparse selection from the many publications about Charles Darwin.

The place of publication is London, unless otherwise stated. If a title is not self-explanatory, I have added a comment.

[Anonymous] *An Elegy on the Much-lamented Death of a most Ingenious Young Gentleman* . . . [Charles Darwin the elder]. G. Robinson, 1778. Written mostly by Erasmus.

Margaret Ashmun, *The Singing Swan*. New York: Greenwood Press, 1968 (First published, 1931). Life of Anna Seward.

John Bowlby, *Charles Darwin: a biography*. Hutchinson, 1990.

Desmond Clarke, *The Ingenious Mr Edgeworth*. Oldbourne, 1965.

Ralph Colp, 'Notes on Charles Darwin's *Autobiography*'. *Journ. Hist. Biol.*, Vol.18, pp.357–401, 1985.

Ralph Colp, 'The Relationship of Charles Darwin to the Ideas of his Grandfather, Dr Erasmus Darwin'. *Biography*, Vol. 9, pp.1–24, 1986.

The Correspondence of Charles Darwin, 1821–1882, A Calendar of. Cambridge: Cambridge University Press, 1994. Brief summaries of some 14 000 letters to and from C.D., with indexes.

The Correspondence of Charles Darwin, Volumes 1–12, 1821–1864. Cambridge: Cambridge University Press, 1985–2001. A total of 32 volumes is scheduled.

Maxwell Craven, *John Whitehurst of Derby*. Ashbourne: Mayfield Books, 1996.

C. D. Darlington, *Darwin's Place in History*. Oxford: Blackwell, 1959. Discusses proponents of evolution from 1790 to 1860.

Charles Darwin [the elder], *Experiments establishing a Criterion between Mucaginous and Purulent Matter*, Cadell, 1780. Includes short biography by Erasmus.

Charles Darwin, *Autobiography* (ed. Nora Barlow). Collins, 1958. Has useful Appendices and Notes. (Also Penguin Classics, 2002.)

Charles Darwin, *On the Origin of Species by Means of Natural Selection*. First ed., Murray, 1859. Facsimile reprint by Harvard University Press, 1964.

[Erasmus Darwin] *A System of Vegetables* . . . translated from the thirteenth edition of the 'Systema Vegetabilium' of . . . Linneus . . . by a Botanical Society at Lichfield. 2 vols. Lichfield: J. Jackson, for Leigh and Sotheby, London, 1783.

[Erasmus Darwin] *The Families of Plants* . . . translated from the last edition of the 'Genera Plantarum' . . . by a Botanical Society at Lichfield. 2 vols. Lichfield: J. Jackson, for J. Johnson, London, 1787.

[Erasmus Darwin] *The Botanic Garden; a Poem, in Two Parts*. Part I, *The Economy of Vegetation*. J. Johnson, 1791 [actually appeared in 1792]. Part II, *The Loves of the Plants*. Lichfield, J. Jackson, for J. Johnson, London, 1789. Facsimile reprints by Scolar Press, Menston, 1973; by Scholarly Press, MI, 1977; by Garland, New York, 1979; by Woodstock Books, 1991 (Part II only).

Erasmus Darwin, *Zoonomia; or, The Laws of Organic Life*. Part I, J. Johnson, 1794 (Vol.1). Parts II and III, J. Johnson, 1796 (Vol.2), issued with 2nd edn. of Vol.1, corrected. Facsimile reprints by A. M. S. Press, New York, 1974; and by Scholarly Press, MI, 1977.

Erasmus Darwin, *A Plan for the Conduct of Female Education in Boarding Schools*. Derby: J. Drewry, for J. Johnson, London, 1797.

Facsimile reprints by Johnson Reprint Corp., New York, 1974; by Routledge, London, 1996, in History of British Educational Thought; by Woodstock Books, 2001.

Erasmus Darwin, Phytologia: or the Philosophy of Agriculture and Gardening. J. Johnson, 1800.

Erasmus Darwin, The Temple of Nature; or, The Origin of Society. J. Johnson, 1803.
Facsimile reprints by Scolar Press, Menston, 1973; and by Scholarly Press, MI, 1977.

Erasmus Darwin, Cosmologia [The poems The Botanic Garden and The Temple of Nature, without the Notes, newly edited by Stuart Harris]. Published by S. Harris, 171 Millhouses Lane, Sheffield S7 2HD. 2002.

Erasmus Darwin, The Letters of Erasmus Darwin (ed. D. King-Hele). Cambridge: Cambridge University Press, 1981.

Francis Darwin (ed.), Life and Letters of Charles Darwin. 3 vols. Murray, 1887.

Francis Darwin and A. C. Seward (ed.), More Letters of Charles Darwin. 2 vols. Murray, 1903.

R. L. and Maria Edgeworth, Memoirs of Richard Lovell Edgeworth. 2 vols. Hunter, 1820.
Facsimile reprint by Irish University Press, Shannon, Ireland, 1969.

R. B. Freeman, Darwin Pedigrees. Printed for the author, 1984. Valuable; but beware of errors in the twentieth-century part of the text.

Donald M. Hassler, The Comedian as the letter D: Erasmus Darwin's Comic Materialism. The Hague: Nijhoff, 1973.

Donald M. Hassler, Erasmus Darwin. New York: Twayne, 1973. Study of his writings.

Edna Healey, Emma Darwin. Headline, 2001. Biography of Charles Darwin's wife – and their extended family.

Randal Keynes, Annie's Box: Charles Darwin, his Daughter and Human Evolution. Fourth Estate, 2001.

Desmond King-Hele, Erasmus Darwin and the Romantic Poets. Macmillan, 1986.

Desmond King-Hele, Erasmus Darwin: a Life of Unequalled Achievement. De la Mare, 1999. (Abbreviated here in the Notes as '1999 Life'.)

Henrietta Litchfield (ed.), *Emma Darwin: a Century of Family Letters*.
  2 vols. Privately printed, C.U.P., 1904. Also Murray, 1915.
J. V. Logan, *The Poetry and Aesthetics of Erasmus Darwin*. Princeton, NJ:
  Princeton University Press, 1936.
Maureen McNeil, *Under the Banner of Science: Erasmus Darwin and his Age*.
  Manchester: Manchester University Press, 1987.
E. Meteyard, *The Life of Josiah Wedgwood*. 2 vols. Hurst and Blackett,
  1865–6.
  Facsimile reprint by Josiah Wedgwood and Sons, Barlaston, 1980.
J. L. Moilliet and B. M. D. Smith, 'A Mighty Chemist': *James Keir*. Privately
  printed, 1983.
Benedict Nicholson, *Joseph Wright of Derby*. 2 vols. Routledge, 1968.
Karl Pearson, *The Life, Letters and Labours of Francis Galton*. 4 vols.
  Cambridge: Cambridge University Press, 1914–30.
Roy Porter, *Enlightenment*. Allen Lane, The Penguin Press, 2000.
Gwen Raverat, *Period Piece*. Faber, 1952. A delightful picture of the
  Darwin family.
L. T. C. Rolt, *James Watt*. Batsford, 1962.
P. Rowland, *The Life and Times of Thomas Day*. Lewiston, NY: Mellen,
  1996.
R. E. Schofield, *The Lunar Society of Birmingham*. Oxford: Oxford
  University Press, 1963.
R. F. Scott, *Admissions to the College of St John the Evangelist in the University
  of Cambridge*, Part III. Cambridge: Deighton Bell, 1903.
Anna Seward, *Memoirs of the Life of Dr Darwin*. Johnson, 1804.
Anna Seward, *The Letters of Anna Seward*. 6 vols. Edinburgh: Constable,
  1811.
W. B. Stevens, *The Journal of the Rev. W. B. Stevens* (ed. G. Galbraith).
  Oxford: Clarendon Press, 1965.
Jenny Uglow, *The Lunar Men*. Faber, 2002. Particularly good on Boulton,
  Watt and Wedgwood.
Barbara and Hensleigh Wedgwood, *The Wedgwood Circle, 1730–1897*.
  Cassell, 1980.
Josiah Wedgwood, *Letters of Josiah Wedgwood* (ed. K. E. Farrer). 3 vols.
  Manchester: E. J. Morton, 1974.

Appendix D
Outline of Ernst Krause's Essay
'The Scientific Works of Erasmus Darwin'
(pages 131–216 of the 1879 book)

Krause begins his essay well (pp.131–2):

> On the second page of the later editions of Darwin's 'Origin of
> Species' (sixth edition, p. xiv, note) we find the following brief
> observation: "It is curious how largely my grandfather,
> Dr. Erasmus Darwin, anticipated the views and erroneous grounds
> of opinion of Lamarck in his 'Zoonomia' (vol.i., pp.500–510),
> published in 1794." Being quite aware of the reticence and
> modesty with which the author expresses himself, especially in
> speaking *pro domo*, I thought immediately that here we ought to
> read between the lines, and that this ancestor of his must certainly
> deserve considerable credit in connection with the history of the
> Darwinian theory. As no light was to be obtained upon this subject
> from German literature, I procured the works of Erasmus Darwin,
> and have found singular pleasure in their study.
>
> I was speedily convinced that this man, equally eminent as
> philanthropist, physician, naturalist, philosopher, and poet, is far
> less known and valued by posterity than he deserves, in
> comparison with other persons who occupy a similar rank. It is
> true that what is perhaps the most important of his many-sided
> endowments, namely his broad view of the philosophy of nature,
> was not intelligible to his contemporaries; it is only now, after the
> lapse of a hundred years, that by the labours of one of his
> descendants we are in a position to estimate at its true value the
> wonderful perceptivity, amounting almost to divination, that he

displayed in the domain of biology. For in him we find the same indefatigable spirit of research, and almost the same biological tendency, as in his grandson; and we might not without justice, assert that the latter has succeeded to an intellectual inheritance, and carried out a programme sketched forth and left behind by his grandfather.

Krause then points out that 'almost every single work of the younger Darwin may be paralleled by at least a chapter in the works of his ancestor'. However, the two Darwins differ 'in their interpretation of nature'.

The elder Darwin was a Lamarckian, or, more properly, Jean Lamarck was a Darwinian of the older school, for he has only carried out further the ideas of Erasmus Darwin, although with great acumen; and it is to [Erasmus] Darwin therefore that the credit is due of having first established a complete system of the theory of evolution. . . . [Yet] we have not the smallest ground for depreciating the labours of the man who has shed a new lustre upon the name of his grandfather. It is one thing to establish hypotheses and theories out of the fulness of one's fancy, even when supported by a very considerable knowledge of nature, and another to demonstrate them by an enormous number of facts, and carry them to such a degree of probability as to satisfy those most capable of judging. Dr. Erasmus Darwin could not satisfy his contemporaries with his physio-philosophical ideas; he was a century ahead of them, and was in consequence obliged to put up with seeing people shrug their shoulders when they spoke of his wild and eccentric fancies (pp.133–4).

Krause then notes the similarity between Erasmus Darwin and Goethe, who also thought 'that a common organization must bind together the higher animals' (p.136).

Next (pp.137–63) Krause discusses Darwin's poem *The Botanic Garden*, beginning with Part I, *The Economy of Vegetation*. He first contradicts the suggestion that the poem may be in imitation of Henry Brooke's *Universal Beauty* (1735) or Sir Richard Blackmore's *Creation* (1712). 'Neither statement has the slightest foundation'. Krause then explains that the four cantos of Part I have as their themes Fire, Earth, Water and Air, and that Darwin addresses 'the nymphs of primeval fire', the 'gnomes or earth-spirits', the 'water nymphs', and the sylphs of air. He quotes about 40 lines of verse, including the verses on the power of steam, 'Soon shall thy arm, Unconquer'd Steam . . .' (p.142). But Krause concentrates on topics that particularly interest him, such as Darwin's discussion of rudimentary organs; and there is a long digression (pp.147–51) about Buffon. This includes a subsidiary digression denouncing the authors of twelve books published between 1711 and 1772 for propounding the 'shallow, sickly enthusiasm which was called "natural religion" ' – the belief that all creatures exist for the service of man. 'Buffon could not escape from this tendency of his time'; but Darwin was not infected by it.

Krause's treatment of Part I is very scrappy, and consists mainly in picking out favoured paragraphs in the Notes. The same applies with Part II, *The Loves of the Plants*: for example, he quotes (pp.154–5) Darwin's discussion of prickles and thorns, concluding with his observation that the large hollies in Needwood Forest

> are armed with thorny leaves [up to] about eight feet high, and have smooth leaves above; as if they were conscious that horses and cattle could not reach their upper branches.

After quoting Darwin on insectivorous plants with approval, Krause shows he has teeth: he savages Darwin for his serious error 'with respect to the secretion of honey in flowers'. Darwin believed

> that plants were generally equipped so as to keep insects and other lovers of honey away from the flowers; and he was strengthened in this opinion by the circumstance that the source of honey in most

flowers is very much concealed, and often hidden under complex protective contrivances (p.158).

This led Darwin to the mistaken belief that flowers 'are as far as possible adapted for self-fertilization'. Darwin 'afterwards wrote impressively upon the mischief of inbreeding'; if only he had heeded 'the magic words "Benefits of Cross-fertilization", his error would have fallen like scales from his eyes' (p.161).

However, Krause praises Darwin for expounding the principle of mimicry 'perhaps for the first time' (p.159), and quotes his analysis of protective coloration (pp.162–3) with approval.

Krause now turns to Zoonomia, 'his chief scientific work', which

essentially forms a physiology and psychology of man as a foundation for a pathology; but at the same time glances are everywhere cast over the whole animal world (p.164).

Krause sees Darwin as believing in 'a living force' that enables plants and animals to adapt 'to the circumstances of the outer world' (p.165): he expands upon this for several pages.

Krause fully discusses (pp.170–91) the evolutionary Section 39 of Zoonomia. He notes how Darwin ridicules the theory of preformation – the idea that all the future progeny existed in miniature inside the animal originally created. Then, in pages 173–84, Krause indulges in a marathon quotation from Zoonomia (vol.i., pages 504–8, with a few gaps), covering Darwin's presentation of the evolution of animals (to use modern parlance).

I am now obliged to summarize this eleven-page quotation. First, Darwin gives reasons for believing that species do change. He points to the changes in form of individual animals during their lives, 'as in the production of the butterfly with painted wings from the crawling caterpillar; or of the respiring frog from the subnatant tadpole' (p.504). So changes in form do occur in nature. Darwin notes the crucial point that monstrosities – or mutations as we should say – may be inherited: 'Many of these enormities of shape are propagated, and continued as a variety at least, if not as a new species of animal' (p.505). These anatomical changes, and the similar structures of the warm-blooded animals, lead him to suggest that all animals may have 'a similar living filament' as a common microscopic ancestor (the theory of common descent, as it is now called). What controls the evolutionary changes? 'The three great objects of desire, which changed the forms of many animals by their exertions to gratify them, are those of lust, hunger and security' (p.506). In considering lust, Darwin looks at those animals where the males combat each other for the exclusive possession of the females. The outcome is 'that the strongest and most active animal should propagate the species, which should thence become improved' (p.507). As Krause remarks later (p.190):

> With the most perfect certainty we [here] have the principles of a theory of sexual selection laid down.

Hunger is the second of Darwin's controlling forces, and he explains how each animal is adapted to its method of acquiring food, citing the trunks of elephants, the rough tongues of cattle and the varied beaks of birds. The third controlling force, security, evokes a discussion of topics like mimicry and protective coloration, as well as the more obvious qualities, such as fleetness of foot, hard shells, etc. Krause points out in a long footnote (p.182) that Darwin does not specify natural selection as the mechanism by which these qualities arise. Krause's quotation from Zoonomia ends with Darwin's lengthy rhetorical question:

> Would it be too bold to imagine, that in the great length of time since the earth began to exist, perhaps millions of ages before the commencement of the history of mankind, would it be too bold to imagine, that all warm-blooded animals have arisen from one living filament ... possessing the faculty of continuing to improve by its own inherent activity, and of delivering down those improvements by generation to its posterity, world without end! (p.509).

Krause then gives two further pages of quotation to show Darwin's views on the evolution of vegetables, including their 'perpetual contest for light and air above ground, and for food and moisture beneath the soil' (p.185). Krause offers three more pages of quotations, including Darwin's acceptance of Hume's idea 'that the world itself might have been generated rather than created' (pp.188–9). With another quotation in support, Krause reminds us that 'Darwin regards sexual reproduction as a principal condition of the advancement of living creatures' (p.190).

Perhaps exhausted by Zoonomia, Krause dismisses Phytologia, Darwin's best scientific book. 'We need not go into it in detail, as his conception of the vegetable world has already been sufficiently explained' (p.192). This leaves a gaping hole in Krause's essay; Darwin's formulation of photosynthesis, his recognition of plant nutrient elements, and dozens of other good ideas are never mentioned.

Krause makes up for this lapse with a long and sympathetic review of The Temple of Nature (pp.192–206). The first canto is entitled 'The Origin of Life', and Krause emphasizes Darwin's beliefs that life arose by spontaneous generation and was in its early stages asexual. Krause goes on:

> This first life originated in the "shoreless" sea:
>
>> Organic life beneath the shoreless waves
>> Was born, and nurs'd in ocean's pearly caves;
>> First forms minute, unseen by spheric glass,
>> Move on the mud, or pierce the watery mass;
>> These, as successive generations bloom,
>> New powers acquire, and larger limbs assume;
>> Whence countless groups of vegetation spring,
>> And breathing realms of fin, and feet, and wing.
>
> In the continuation of these verses (lines 295–302) the author recalls to mind that the higher animals, and even "the image of God", commence their course of life as microscopic creatures and points:

Imperious man, who rules the bestial crowd,
Of language, reason, and reflection proud,
With brow erect who scorns this earthy sod,
And styles himself the image of his God,
Arose from rudiments of form and sense,
An embryon point, or microscopic ens!

Then, when mountains upheaved by the central fire, or coral reefs, first rose above the surface of the boundless sea, individual living organisms landed upon them, and passing through an amphibious condition, became aerial creatures. "After islands or continents were raised above the primeval ocean", he says, in a note on p.29, "great numbers of the most simple animals would attempt to seek food at the edges or shores of the new land, and might thence gradually become amphibious; as is now seen in the frog, who changes from an aquatic animal to an amphibious one; and in the gnat, which changes from a natant to a volant state. . . . Those [organisms] situated on dry land and immersed in dry air, may gradually acquire new powers to preserve their existence; and by innumerable successive reproductions for some thousands, or perhaps millions of ages, may at length have produced many of the vegetable and animal inhabitants which people the earth."

In the second canto, entitled 'Reproduction of Life', Krause quotes only from the Notes, to show how Darwin recognized 'the mischief of inbreeding'.

As the sexual progeny of vegetables are thus less liable to hereditary diseases than the solitary progenies, so it is reasonable to conclude, that the sexual progenies of animals may be less liable to hereditary diseases, if the marriages are into different families, than if into the same family; this has long been supposed to be true, by those who breed animals for sale. (Additional Notes, p.44.)

After this, Krause mentions the warning against marrying heiresses, already quoted by Charles Darwin in the main text.

In the third canto, 'Progress of the Mind', Krause picks out two processes that Darwin discussed. The first is the importance of the hand, and its improved sense of touch in humans, in enabling the human mind to improve. The second is the human love of imitation, which Darwin personifies as the 'Muse of Mimicry' and sees as 'the first origin of all moral actions, languages and arts' (p.201).

In the fourth canto, 'Of Good and Evil', Krause is attracted by Darwin's 'description of the pitiless struggle for existence, which rages in the air, on the earth, and in the water':

> Air, earth, and ocean, to astonish'd day
> One scene of blood, one mighty tomb display!
> From Hunger's arm the shafts of Death are hurl'd,
> And one great Slaughter-house the warring world! (lines 63–66)

Not only do animals destroy each other and plants, but even the plants struggle among themselves for soil, moisture, air, and light:

> Yes! smiling Flora drives her armed car
> Through the thick ranks of vegetable war;
> Herb, shrub, and tree, with strong emotions rise
> For light and air, and battle in the skies;
> Whose roots diverging with opposing toil
> Contend below for moisture and for soil;
> Round the tall Elm the flattering Ivies bend,
> And strangle, as they clasp, their struggling friend;
> Envenom'd dews from Mancinella flow,
> And scald with caustic touch the tribes below;
> Dense shadowy leaves on stems aspiring borne
> With blight and mildew thin the realms of corn;
> And insect hordes with restless tooth devour
> The unfolded bud, and pierce the ravell'd flower. (lines 41–54)

... Nevertheless ... without such a struggle, nearly every creature would soon overrun the whole world:

All these, increasing by successive birth,
Would each o'erpeople ocean, air, and earth.

Darwin's idea of organic happiness, based on his belief that many mountains are the remains of living creatures, inspires the last of Krause's quotations (p.206):

Thus the tall mountains, that emboss the lands,
Huge isles of rock, and continents of sands,
Whose dim extent eludes the inquiring sight,
ARE MIGHTY MONUMENTS OF PAST DELIGHT;
Shout round the globe, how Reproduction strives
With vanquish'd Death,—and Happiness survives. (lines 447–52)

Krause believes (pp.207–8) that Darwin

directed the eyes of many of his readers to the struggle for existence, and in this we may perhaps find the explanation of the remarkable fact that so many English naturalists (Wells, Matthew, Charles Darwin, Wallace, &c) have one after the other set up the principle of natural selection.

Looking back on Darwin's poems, Krause wonders why they are so little appreciated, and he quotes at some length the judgment by G. L. Craik in his *History of English Literature* (1868). Craik offers a mixture of censure and praise:

Nothing is done in passion and power ... Every line is as elaborately polished and sharpened as a lancet ... No writer has surpassed him in the luminous representation of visible objects in verse. (pp.209–10)

Krause's own estimate is kinder (p.210):

We will be more just, and say, that since the time of Lucretius, hardly any attempt to combine the opposing spheres of science and poetry in a didactic poem, and to put forth therein entire systems, has been so successful as in Darwin's works.

In summarizing Darwin's scientific work, Krause says (p 211):

We must in the first place admit that *he was the first who proposed and consistently carried out, a well-rounded theory with regard to the development of the living world*, a merit which shines forth most brilliantly when we compare with it the vacillating and confused attempts of Buffon, Linnæus and Goethe.

His scheme was much better than the old 'comparison of Nature with a great piece of clockwork'; but Darwin's 'extension of his theory to the vegetable kingdom has robbed it of the efficacy which it might have attained if limited to the animal kingdom', Krause believes.

There is a curious final paragraph (p.216), which begins as an appropriate rounding-off, and then suddenly changes in tone and subject:

Erasmus Darwin's system was in itself a most significant first step in the path of knowledge which his grandson has opened up for us; but to wish to revive it at the present day, as has actually been seriously attempted, shows a weakness of thought and a mental anachronism which no one can envy.

It is the last half of this sentence that was seized on by Samuel Butler as an underhand attempt to rubbish his book *Evolution, Old and New*.

Appendix E

Linking Page Numbers in the 1879 Book with Those in the Proofs

The proofs were cut up and re-assembled by Henrietta, and then pruned and added to. So, given any page in the book, which pages in the proofs does it stem from? Or are parts of it not in the proofs at all? There are no logical answers; so I have provided the list below, in which the minimum 'recognized' length of text is 2 lines, and the symbol N denotes 'not in the proofs'.

1879 book page	pages in proofs	1879 book page	pages in proofs	1879 book page	pages in proofs
1	2, N	17	15	33	36, 37, 41
2	N, 2, 3	18	16	34	41, 62
3:	picture	19	16	35	62, 63
4	3	20	17	36	63, 48
5	4	21	17, 18	37	39
6	9, 10	22	18	38	39, 40
7	10, 11	23	18, 19	39	40
8	11	24	19	40	40, 42, 44, 42
9	11, 12	25	19	41	42, 43
10	12	26	20, 19, N, 65	42	43
11	12, 13	27	19, N, 34	43	43, 46
12	13	28	34, 35	44	46, 62, 48
13	13, 14	29	35, 64	45	48, 42
14	14, 3	30	64, 35	46	42, 68
15	3, 14, 15	31	35, 36	47	61, 68, N
16	14, 15	32	36	48	N, 44

1879 book page	pages in proofs	1879 book page	pages in proofs	1879 book page	pages in proofs
49	44, 45	76	58	103	24, 25
50	45, 41	77	58, 59	104	25
51	41, 45	78	59, 60	105	25, 26
52	43, 46	79	60	106	26, 27
53	46, 69	80	61, 4	107	27, N
54	69, 66	81	5	108	27
55	66, N	82	5	109	27, 28
56	N, 67	83	58, 5, 6	110	28
57	67	84	6	111	28, 29
58	N, 67	85	7	112	29
59	68	86	7, 8	113	29, 47
60	63	87	8, 9	114	47, 30
61	64, 63	88	9, 24, 20, 48–9	115	30
62	63, 54, 65	89	21	116	30, 31
63	65	90	21, 49	117	31
64	65, 66	91	49, 50	118	31, 32
65	66, 50, 51	92	50, 21, 22	119	32, N
66	51	93	22, N	120	N, 32
67	52	94	22	121	32, 33
68	53	95	23	122	33
69	53	96	23, N	123	33, 34
70	54	97	23, 24, 37	124	34, 70
71	55	98	37	125:	picture
72	55	99	37, 38	126	70, 71
73	56	100	38	127	71
74	56, 57	101	38, 39, 24		
75	57	102	24		

Acknowledgements

This unabridged edition of Charles Darwin's book relies on the first proofs, as corrected by Charles and Henrietta before they decided on abridgement. I am grateful to the late George Pember Darwin and the Syndics of the Cambridge University Library for permission to use the proofs, which are in the Darwin Archive at the Library (DAR 210.11:45).

I have much appreciated the encouragement of Patrick Zutshi, Keeper of Manuscripts at the Library, and the essential help of Godfrey Waller and the staff of the Manuscripts Room.

I am greatly indebted to Valerie McMillan, who received my amended version of the 1879 proofs, with the Introduction, Notes and Appendices, and magically transformed them into the disc from which the book was printed.

I would like to thank Rosemary Thomas for her perceptive and constructive comments on the Introduction; Hugh Torrens, for knowing so much that no one else knows; Mike Crump and Anne Summers of the British Library; and many others who kindly helped with detailed information, including Gordon Cook, Maxwell Craven, Angela Darwin, David Fraser, Nick Gill, Edna Healey and Nick Redman.

For permission to reproduce illustrations, I am grateful to William Darwin and the Darwin Heirloom Trust (for the frontispiece); and to Rosemary Bonham-Smith (for the drawing of Breadsall Priory).

At the Cambridge University Press, I thank Simon Mitton for his belief in the book, and Shana Coates, who has guided it along the perilous path of production so swiftly, skilfully and pleasantly.

Index

As well as the main text, this index covers the Introduction, the Notes and the outline of Krause's essay. It does *not* cover Appendices A, B, C or E, nor books mentioned only in the Notes. Dates of birth and death are given for people born between 1600 and 1850. For Darwin family members, their relationship to Erasmus is given in parentheses. Erasmus's books are indexed under their titles. Figures in **bold type** indicate leading references.

508
.092
Darwi
-D

Darwin, C.
Charles Darwin's the life of
Erasmus Darwin.
Aurora P.L. SEP12
33164004720384